Resources, Social and Cultural Sustainabilities in the Arctic

This book focuses on the understudied social and cultural dimensions of sustainability in the Arctic. More specifically, it explores these thematics through paying attention to resources in different definitions and forms and the ways in which they entangle in the realities and expectations of social and cultural sustainability in the region.

The book approaches resources as socially and culturally constructed and also draws attention to social, human and cultural capabilities and the roles they have in making and shaping the imaginaries of sustainability. Together, this volume and its case studies contribute to a broadened understanding of the interplay of natural and material resources and social and cultural capabilities as well as their discursive framings.

This multidisciplinary text includes contributions from political sciences, sociology, gender studies, regional studies, economics and art research. With its wide range of conceptually informed case studies, the book is relevant for researchers and professionals as well as advanced students and for institutions and organizations offering education in Arctic affairs.

Monica Tennberg, a research professor, conducts research about Arctic political economy: that is, about connections between wealth, power and well-being. She has recently contributed to the book *Adaptation Actions for a Changing Arctic – Perspectives from the Barents Area* (2017).

Hanna Lempinen is a university lecturer in political science at the University of Lapland, Finland, and a visiting senior researcher at the Arctic Centre, University of Lapland. Her research interests include social and cultural aspects of sustainability, especially in the context of Arctic large-scale energy and industrial development. Her book *Arctic Energy and Social Sustainability* was published by Palgrave in 2018.

Susanna Pirnes is a doctoral candidate in political science. Her research interests are related to the Russian Arctic, Arctic identities, memories and history politics.

Routledge Research in Polar Regions
Series Editor: Timothy Heleniak
Nordregio International Research Centre, Sweden

The Routledge series in Polar Regions seeks to include research and policy debates about trends and events taking place in two important world regions: the Arctic and Antarctic. Previously neglected periphery regions, with climate change, resource development and shifting geopolitics, these regions are becoming increasingly crucial to happenings outside these regions. At the same time, the economies, societies and natural environments of the Arctic are undergoing rapid change. This series seeks to draw upon fieldwork, satellite observations, archival studies and other research methods which inform about crucial developments in the Polar regions. It is interdisciplinary, drawing on the work from the social sciences and humanities, bringing together cutting-edge research in the Polar regions with the policy implications.

Arctic Sustainability Research
Past, Present and Future
Andrey N. Petrov, Shauna BurnSilver, F. Stuart Chapin III, Gail Fondahl, Jessica Graybill, Kathrin Keil, Annika E. Nilsson, Rudolf Riedlsperger, and Peter Schweitzer

Resources and Sustainable Development in the Arctic
Edited by Chris Southcott, Frances Abele, Dave Natcher, and Brenda Parlee

Performing Arctic Sovereignty
Policy and Visual Narratives
Corine Wood-Donnelly

Resources, Social and Cultural Sustainabilities in the Arctic
Edited by Monica Tennberg, Hanna Lempinen, and Susanna Pirnes

For more information about this series, please visit: www.routledge.com/ Routledge-Research-in-Polar-Regions/book-series/RRPS

Resources, Social and Cultural Sustainabilities in the Arctic

Edited by
Monica Tennberg, Hanna Lempinen
and Susanna Pirnes

Routledge
Taylor & Francis Group

LONDON AND NEW YORK

First published 2020
by Routledge
2 Park Square, Milton Park, Abingdon, Oxon OX14 4RN

and by Routledge
52 Vanderbilt Avenue, New York, NY 10017

First issued in paperback 2021

Routledge is an imprint of the Taylor & Francis Group, an informa business

British Library Cataloguing-in-Publication Data
A catalogue record for this book is available from the British Library

Library of Congress Cataloging-in-Publication Data
A catalog record for this book has been requested

ISBN: 978-0-367-17544-3 (hbk)
ISBN: 978-1-03-208787-0 (pbk)
ISBN: 978-0-429-05736-6 (ebk)

Typeset in Times New Roman
by Apex CoVantage, LLC

Contents

List of figures   vii
List of maps   viii
List of tables   ix
List of contributors   x
Acknowledgments   xi

1 Sustainabilities in the resourceful North   1
  HANNA LEMPINEN, MONICA TENNBERG AND SUSANNA PIRNES

PART I
Entangled resources and sustainabilities   13

2 Greenland and the elusive better future: the affective
  merging of resources and independence   15
  MARJO LINDROTH

3 Promise and threat: living with nuclear in the Finnish context   27
  HANNAH STRAUSS-MAZZULLO

4 Untied resource as a threa/-t/-d for social fabric(ation)   41
  JOONAS VOLA

5 "Prudent development": the (r)evolution of the Arctic
  energy concern in the 2007–2017 Arctic Energy Summit Reports   56
  HANNA LEMPINEN

6 Socially responsible investments (SRIs) in the European
  Arctic: new pathways for global investors to outperform
  conventional capital investments?   69
  ADRIAN BRAUN

7 Resources on the Arctic border: views of the Finnish
   municipalities and the EU's cross-border program   83
   PAULA TULPPO

PART II
Whose imaginaries?   101

8 The political ecology of Northern adaptation: power, nature
   and knowledge   103
   GEMMA HOLT

9 Arctic expertise and its social dimensions in Lapland   117
   MONICA TENNBERG

10 When gender matters: equality as a source
   of Arctic sustainability?   131
   HEIDI SINEVAARA-NISKANEN

11 Sámi cultural heritage and tourism in Finland   144
   FRANCIS JOY

12 History as a resource in Russian Arctic politics   163
   SUSANNA PIRNES

13 The resourceful North: divergent imaginaries from the
   European Arctic   175
   MONICA TENNBERG, HANNA LEMPINEN AND SUSANNA PIRNES

   Index   182

Figures

6.1   Purposes for raising capital in the framework of socially
      responsible investments (e.g. by issuing climate bonds or
      sustainability bonds or establishing ESG funds)   71
11.1 A Sámi drum that originated from somewhere within the Kemi
      Sámi area but has an otherwise uncertain Sámi origin   149
11.2 An artificial representation of a constructed resource for the
      development of tourism and a marketing tool: two fake Sámi
      drums exhibited in a souvenir shop in Ivalo   152
11.3 A plastic tray containing a combination of drum landscapes
      that have been copied and modified from the sacred *noaidi*
      drum from Finland. The tray was on display in a souvenir
      shop in Ivalo in Finnish Sápmi. These illustrations are sources
      of traditional knowledge for the Sámi and are therefore
      considered their common property. These designs also appear
      on wallpaper and curtains   158

Maps

3.1 Actual and future nuclear power production sites in Finland as
    well as selected towns included in the assessment process   30
4.1 Kemi River, Finnish Kemijoki   43
7.1 The Arctic border region between Finland and Sweden   85
7.2 A map of the Interreg V A Nord program region (Interreg Nord, 2018) 89

Tables

8.1 The policy goals of the AACA vary in their potential to integrate multiple knowledges 113

9.1 EU social progress index for Northern and Eastern Finland and Upper Norrland in Sweden, 2011 121

9.2 Regional potential index by region, 2015 122

Contributors

Adrian Braun is a doctoral candidate at the Research Centre for Socially and Environmentally Responsible Mining, University of Eastern Finland.

Gemma Holt was a US Fulbright student (2017–2018) at the Arctic Centre, University of Lapland, Finland. She is a nonprofit strategy consultant at TDC in Boston, MA, USA.

Francis Joy is a postdoctoral researcher in the Faculty of Arts and Design, University of Lapland, Finland.

Hanna Lempinen is a university lecturer in the Faculty of Social Sciences, University of Lapland, Finland.

Marjo Lindroth is a university researcher at the Arctic Centre, University of Lapland, Finland.

Susanna Pirnes is a doctoral candidate at the Arctic Centre, University of Lapland, Finland.

Heidi Sinevaara-Niskanen is a university researcher in gender studies at the University of Lapland, Finland.

Hannah Strauss-Mazzullo is a postdoctoral researcher at the Arctic Centre, University of Lapland, Finland.

Monica Tennberg is a research professor at the Arctic Centre, University of Lapland, Finland.

Paula Tulppo is a doctoral candidate at the Arctic Centre, University of Lapland, Finland.

Joonas Vola is a junior researcher in the Faculty of Social Sciences, University of Lapland, Finland.

The book was written as a collaboration by the members of the Northern political economy research group at the Arctic Centre, University of Lapland.

Acknowledgments

We wish to thank the anonymous chapter reviewers for valuable comments and constructive criticism. We also express our gratitude to Pirkko Hautamäki, our language expert. All remaining errors are our responsibility.

The fieldwork and the writing of chapter 3 by Hannah Strauss-Mazzullo was funded by the Kone Foundation (decision number 085192, 5.12.2014). Marjo Lindroth's research for chapter 2 is part of the project *Indigeneity in Waiting: Elusive Rights and the Power of Hope*, funded by the Kone Foundation and the Academy of Finland (decision number 295557). Heidi Sinevaara-Niskanen's chapter 10 is part of the project *Indigeneity in Waiting: Elusive Rights and the Power of Hope*, funded by the Academy of Finland (decision number 295557). The Fulbright Finland Foundation supported the research and writing of chapter 8 by Gemma Holt.

1 Sustainabilities in the resourceful North

Hanna Lempinen, Monica Tennberg and Susanna Pirnes

A small, remote town in Eastern Finland

In the midst of complex, interlinked climatic and environmental changes and their multiple consequences in the Arctic, people and their communities ideally use their resources rationally, strive for economic diversification, have a strong collective identity and support public participation to cope with and adapt to these changes and transform themselves to maintain their sustainability (a sustainable local community, see Raco, 2005). An illustrative example of contemporary challenges posed to people and their communities in the European Arctic can be found in Salla, a municipality near the Finnish-Russian border in Eastern Finland. It is a small, remote municipality and, in many ways, it is also in decline, tackling as it is the challenges of high unemployment, declining livelihoods, outmigration and an aging population. These trends raise concerns for the future of the municipality and its ability to provide basic services and infrastructure for its residents. Resources to support and further develop Salla include its geographical location near the Finnish-Russian border "as a gateway" for transportation and tourism; its surrounding northern, unique wilderness; and some existing but rather modest infrastructure for tourism compared to many other ski resorts in Finnish Lapland. Tourism is an important livelihood and employer and almost seems to be the only sector with prospects for future economic development (Tennberg and Lempinen, 2015).

Salla promotes itself as "a place where nothing happens," yet in early 2010" the small municipality found itself "in the middle of everything": plans were being made for visa-free travel between the EU and Russia, and new mines were expected to open in the region. Increased mining, in turn, would have given the impetus to a new rail link between Finland and Russia, which was planned to provide a route to as far as the Arctic Ocean and make the municipality a hub for tourism, industrial development and transportation. Many of these plans, however, were put on hold due to the freezing of EU-Russian relations in the aftermath of the war of Ukraine in 2014, and Salla again found itself at a crossroads with regard to its future. Salla's opportunities to make decisions about its own future, resources and economic diversification are limited. Any expectations relating to the future development of the Salla municipality depend heavily on political and economic actors working elsewhere and decisions made elsewhere, far from the community. The municipality, its people and their livelihoods are therefore highly vulnerable to external changes and disturbances (Tennberg and Lempinen, 2015).

The example of Salla leads us to the problematics of social sustainability and its multiple dimensions, whose study continues to pose theoretical and analytical challenges due to the ambiguous nature of both the notion of "the social" and (its) sustainability (Vallance et al., 2011; Murphy, 2012; Boström, 2012; Latour, 2005). These hitherto largely ignored challenges have recently been considered in the field of Arctic studies where the revival of interest in the societal aspects of sustainability is seen in a growing number of multidisciplinary endeavors to grasp and understand the societal facets of Arctic sustainability and its different definitions, challenges and future trajectories (Petrov et al., 2017; Fondahl and Wilson, 2017; Pram Gad and Strandsbjerg, 2019). They have contributed to the emergence of a new interdisciplinary field of study: Arctic sustainability science (Petrov et al., 2017).

The Arctic and its development have been discussed mostly from the point of view of natural resources and the implications of their extraction for local communities and indigenous peoples. In contrast, Arctic sustainability science has moved away from perceiving the Arctic as a global resource storehouse and toward investigating the complexity of entanglements of resources, local life and social development. This shift in emphasis has been largely motivated by the increasing awareness of the inadequateness of existing indicators to measure societal development and the politicized nature of the concept of sustainable development as well as the pivotal role that human and social sciences can have in conceptualizing and solving Arctic sustainability concerns. Our book is an empirically grounded contribution to these conceptual and theoretical debates and to the discussion focusing on the relationship between resources, development and social and cultural (un)sustainability. The book and its diverse Euro-Arctic case studies focus on and illustrate different aspects of why the Arctic resource discourses and practices as they exist today are unsustainable, drawing attention to often-neglected human, social and cultural aspects of the sustainability debate.

Defining social and cultural sustainability

The notions of sustainable development and sustainability are neither new nor unexplored. The term "sustainable development" first appeared in a 1980 report of the International Union for Conservation of Nature (IUCN), but the underlying concerns over the carrying capacity of ecosystems in the face of increasing human impacts and stressors had already been present(ed) in many of the foundational works associated with the environmental awakening of the 1960s and 1970s (e.g. Carson, 1962; Meadows et al., 1972). Politically, the concept saw its breakthrough after the publication of the Brundtland report *Our Common Future* (WCED, 1987), which defined sustainable development as "development that meets the needs of the present without compromising the ability of future generations to meet their own needs." Sustainable development became the core principle of global environmental cooperation in the 1992 United Nations Conference on Environment and Development. It remains one of the main aims in the Arctic cooperation and has retained its robustness as a policy concept in national and international discourses. Also, the United Nations recently accepted the new 2030 millennium goals for sustainable development (UN, 2015). However, the notions of sustainability and sustainable development have been increasingly acknowledged also as political: their applications arrange and rearrange the inter-relations between different actors and different entities as well as the understanding of a good life today and tomorrow. This makes sustainability "*a concept that does something to the way in which politics unfold*" (Pram Gad et al., 2017, p. 15; emphasis added).

Traditionally, sustainable development has been understood as comprising three interconnected and mutually supportive dimensions or pillars: the economic, environmental and social aspects of development. Of these, the social dimension has broadly been recognized as the most elusive. The social dimension in the sustainability debates and agendas has been characterized as "fluid" or "dismissed altogether" (Boström, 2012, p. 1), and it has widely been acknowledged as "more difficult to analyze, comprehend, define, and incorporate into sustainability projects and planning than the other dimensions of sustainability" (Boström, 2012, p. 6). The messy nature of the debates revolving around the social dimensions of sustainability and sustainable development have led to various interpretations of what the concept might entail. Most commonly, the "social" in the sustainable has been approached and constructed in contextual and *developmental* terms: as concrete development targets and policy goals such as promoting equity, diversity, social cohesion, quality of life, democratic governance and responsibility (Sen, 1999; Murphy, 2012; Partridge, 2014). From this perspective, many signs of socially unsustainable development can be observed both in the broader Arctic region and in terms of our case study on Salla: youth out-migration, aging populations, the precarious nature of employment and various social problems, including increasing numbers of suicides (Tennberg and Lempinen, 2015).

Meanwhile, other interpretations of social sustainability have highlighted the necessity of *procedural* aspects and practices in achieving and maintaining sustainability (Vallance et al., 2011; Del Río and Burguillo, 2008; Vanclay, 2002, 2003). In this, the perceptions and impressions of the individuals and communities affected in one way or another by ongoing development play a role, as does their ability to influence this development. For local people and decision-makers, being able to participate in and influence the development of their region is an important resource as well as a prerequisite for sustainability. Furthermore, communities are not homogenous: even members of the same community can feel very differently about ongoing developments, be unevenly impacted by them and have very different understandings of what social and cultural sustainability or sustainable development might mean (Nuttall, 2010, p. 197; Slootweeg et al., 2001, p. 25; Vanclay, 2003, p. 7). Also, in our example case of Salla, the local residents voiced diverse views, expectations, concerns and hopes regarding the municipality's future. For example, while some unreservedly saw increasing tourism as the developmental strategy for the municipality, others expressed concerns about the impacts that its dominance would have for the prospects of other livelihoods and professionals in the area. Coping with disagreements and conflicts inside a community is also an important aspect of social and cultural sustainability (Tennberg and Lempinen, 2015).

In the discussions of the social aspects of sustainability, culture and related issues have traditionally been implicitly included under the broader umbrella of social sustainability (see e.g. Del Río and Burguillo, 2008; Psaridikou and Szerszynski, 2012; Hiedanpää et al., 2012). However, recent years have witnessed an emerging strand of literature highlighting the necessity of understanding culture as a separate – although closely enmeshed and intertwined – dimension of sustainability. Among these contributions, cultural sustainability has been defined, in terms of cultural heritage, as the vitality of local communities and societies and as cultural changes required to achieve sustainability (Soini and Birkeland, 2014); but attention has also been devoted to what exactly constitutes the "culture" that should be sustained. While no universally applicable definition of what culture entails is either possible or desirable, in its broadest terms, culture can be understood as the diversity of ways living, being and making sense of the world (see e.g. Mercer, 2002; Wilk, 2002; Dessein et al., 2015, p. 31). In this respect, there is not only one "Salla" in our example case but many of them: a diversity of village life and identities in the villages surrounding the municipal center. Despite youth out-migration and an aging population, there is a strong sense of identity and commitment to continuity of common life (Tennberg and Lempinen, 2015).

Our contribution to the Arctic sustainability debate

In this book, we take a focus on the social and cultural dimensions of sustainability in the circumpolar North. More specifically, we explore these thematics through paying attention to resources – in different definitions and forms – and the ways in which they entangle in the realities and expectations of social and cultural sustainability in the case study context of the Arctic region. In doing this, we distance ourselves from the conventional understanding of resources as quantifiable and unquestionable states of the world that can be extracted and harvested and that can be assigned a monetary value that would somehow be accurate, objective and calculable (Lähde, 2015, pp. 60–62; Ferry, 2016). Instead, the chapters of this book approach the question of resources from two separate but interlinked perspectives. On the one hand, (natural) resources and their value and importance are understood as socially and culturally constructed, thus inviting discussion on "how they are constructed, by whom and for whom" (Nilsson and Filimonova, 2013, p. 3; also Bridge, 2009). On the other hand, in other chapters, a conscious effort is made to shift the focus of inquiry away from the natural resources that so often dominate resource-related sustainability debates in the context of the Arctic region and to address issues of social, human and cultural capabilities and the roles they have in making and shaping the landscapes of Arctic sustainability (see also Nuttall, 2002). Together, this volume and its case studies contribute to a broadened understanding of the interplay of natural and material resources and social and cultural capabilities as well as their discursive framings. As a result, the book takes a fresh look at what a "resource" might entail and problematizes the often simplistically applied notion by drawing attention to their ambiguity, unpredictability and imperfection. The contributions of this book will thus advance the understanding of complex relations and dynamics between resources, peoples and development in the North beyond the mainstream studies of Arctic governance and sustainability science. Our findings and observations about these entanglements are not only relevant in the regional context but also contribute to discussions and debates on the contradicting and conflicting dynamics of sustainable development in an increasingly globalized world.

Another aspect of the sustainability debate that we wish to draw attention to and explicate has to do with the crucial differences between (the notions of) sustainable development and sustainability, more often than not used interchangeably despite the marked differences in their fundamental meanings. Whereas *sustainable development* implies either development that does not undermine the basis of its own continuity or, alternatively, development toward a state of sustainability, *sustainability* refers to a state of being or a way of living which can, at least in principle, be maintained indefinitely (Kassel, 2012, p. 34; Missimer et al., 2010, p. 1108). The notion of sustainability thus refers to "ongoing support of life as we know it" (Kassel, 2012, p. 34): that is, maintaining a steady state instead of "developing" or "growing." There is, as critique points out, a mismatch between sustainable development and sustainability. Sustainable development, embracing the idea of continuous growth, even if considered inclusive, equitable and participatory, can still be fundamentally opposed to sustainability in the true sense of the word: creating and maintaining conditions that could be continued into the future without destroying the social and ecological foundations they are built on. In this respect, our volume is a bold critique of the contemporary Arctic sustainability science that aims to apolitically reconcile the oftentimes irreconcilable dimensions of economic, environmental and social "development." Through this emphasis, we draw attention to the culturally bound and value-laden nature of the very notion of sustainability itself. Its underlying assumptions about the nature of the societies and the environment, what constitutes "development" and how best to achieve it are thoroughly culturally mediated. Not only the cultural patterns of acting and thinking that are causing the challenges of sustainability in contemporary societies, but also the solutions and means for addressing them are inherently and unavoidably penetrated by culture (WCED, 1987; Dessein et al., 2015, pp. 29–31). The question arises: are our cultures thinking about and implementing sustainability unsustainably?

Alongside the focus on the interplay of resources and sustainability – and specifically its social and cultural aspects and articulations – the approach of this book underlines the urgent need to address sustainability as the interdisciplinary and multidisciplinary issue that it is. The authors of this volume, all members of the Northern political economy research group at the Arctic Centre, University of Lapland, Finland, investigate and explore social and cultural sustainability in the context of Euro-Arctic case studies through the lenses of and often also combining the perspectives of a wide range of different human and social sciences; equally, anthropology, regional science, economics, gender studies and political sciences are represented on the pages of this book. This emphasis on true interdisciplinarity responds to the calls made within the field of Arctic sustainability science (see Petrov et al., 2017) and in broader social sustainability literature that seek an understanding of the "social" that would not be understood as separate from but as embedded in the other, equally artificial aspects, pillars or dimensions of sustainability (see e.g. Lempinen, 2018, pp. 55–56).

Structure of the book

This volume is divided into two sections: Part 1 takes a broad focus on entanglements of resources and sustainabilities, and Part 2 takes the question of knowledge in relation to sustainability as its core. In the first part, the discussion is opened by Marjo Lindroth, whose chapter critically assesses the entanglement of resource extraction and the pursuit of independence in Greenland. As Greenland continues to be economically dependent on the (former) colonizer Denmark, reaching economic self-sufficiency has become a critical political aim. Both the extraction of resources and independence hold out the promise of a better future, one with increased wealth and equality among nations. This entwinement also creates fears about the potential negative social, environmental and cultural impacts of resource extraction. Lindroth argues that affective elements have a significant role in the debates on Greenland's future. By engaging in a discussion of affective elements that are engendered by these visions for the future, the chapter holds that while Greenland's resources and independence already have a material presence, they remain elusive. Affective elements, as the chapter points out, have a powerful role in contemporary politics in facilitating or preventing certain types of developments.

Hannah Strauss-Mazzullo, whose contribution is based on a case study in Northern Finland, discusses the social acceptance of a new nuclear power plant and a waste storage. The communities in question, Pyhäjoki and (in the southwest of Finland) Eurajoki, generally support the plans while the protest is scattered and met with local opposition. The social acceptance of these projects relies on an entanglement/intertwinement of local and national discursive practices. Locally, the plant provides employment and revenues, and nationally, the plant advances independence from Russian electricity imports and meeting obligations for mitigation of greenhouse gases. While the economic benefits of the nuclear power plant and storage appear short term and tangible, the problems related to nuclear power and waste, such as accidents and catastrophes, are mostly ignored. The successful handling of high-risk nuclear technology since the 1970s continues to be an issue of national pride.

In the following chapter, Joonas Vola concentrates on the potential future development of the Arctic Kemijoki River as a resource in Finnish Lapland. The debate about the river and its use produces an entanglement of regional development, flood risk and economic potentialities of the river. In this discourse, the river could be utilized further for hydropower and to improve the local and regional socioeconomic development, while without such development a major risk for the infrastructure and inhabitants of the city of Rovaniemi may be realized. The current state of the river is described as socially unsustainable due to the waste of its potential as a resource and because of the threat of flooding it causes for the community. The proposed measures in flood protection management for both gaining economic benefits and preventing floods crucially compromise the ecological sustainability and the life of the river and narrow down the societal development to a certain predestined path.

In her chapter, Hanna Lempinen focuses on Arctic energy resources through a case study on the hitherto largely unexplored question of how the regional energy concern has been conceptualized *from and for* the Arctic region. While the Arctic and its energy reserves tend to be predominantly understood in terms of oil and gas exports for international markets, Lempinen examines how the underrepresented and grossly understudied social dimension of sustainability is entangled with the regional energy concern. The chapter also draws attention to the changing yet weighty role that is being envisioned for Arctic energy resources in the rapidly warming and gradually decarbonizing (energy) world.

Next, Adrian Braun discusses innovations in climate policies and their funding. The European Arctic is in the process of transition due to the increasing impacts of climate change and corresponding mitigation and adaptation strategies. The "socially responsible investments" (SRIs) are a potential financial instrument to raise capital to support mitigation of emissions and adaptation measures to climate change. SRIs seek to consider both financial return and the well-being of the society; the aim at a positive social (or environmental) impact needs to be addressed for an investment to be eligible as an SRI. These "green" investments provide opportunities for various actors, such as municipalities, cities and corporations, to contribute to the well-being and societal development in the region. Investments into sustainable infrastructures, green housing and circular economy projects may provide pathways to long-term achievements that strive for social and environmental sustainability but still remain a modestly used opportunity in the European Arctic. While being a fairly novel investment concept in the Arctic context, SRIs already have a strong presence on the big financial markets (e.g. New York, Luxembourg, London), and it will be interesting to see how the Nordic markets in Europe follow up on this development in the next few years.

In the final chapter of the first part, Paula Tulppo discusses the importance of different resources for social and cultural sustainability in the context of Tornio Valley, an area for the EU's cross-border cooperation between Finland and Sweden. The chapter analyzes what resources, entangled or not, for regional development can be identified in the documents of the EU's cross-border cooperation program (Interreg V A Nord, 2014–2020) and those drawn up by the Finnish municipalities in the region. The textual analysis of the EU and municipal documents shows similarities in the environmental, economic, cultural and social resources, but the emphasis by the two actors is different. The resources identified in the municipal strategies relate more closely to a good everyday life while the resources in the EU documents clearly focus more on the economic aspects.

The second part of the volume, exploring the entanglements and intertwinements between different forms of knowledge and producers, holders and users of knowledge, is opened by Gemma Holt. Her chapter focuses on the intersecting roles of knowledge and social values in the scientific assessment of adaptation options to changing environmental conditions from the perspective of political ecology. Her aim is to demonstrate how environmental issues and a changing political landscape have helped shift the meanings and practices of science. The chapter discusses the determination of whose knowledge is relevant for adaptation, how knowledge can be effectively communicated and how it can be linked to policy and practice. The chapter is based on an analysis of the Arctic Council's project Adaptation Actions for a Changing Arctic (AACA), in particular the Barents area report as a case study, to assess how multiple knowledges are used and serve as a basis for theorizing about the politics of adaptation.

In the next chapter, Monica Tennberg investigates the social dimensions of Arctic expertise in Finnish Lapland. Arctic expertise has become a popular approach in European "smart" regional development based on the idea of region-specific skills, competences and knowledge that may lead into new innovations, products and services. Arctic expertise builds not only on a competent, mobile and healthy workforce, but also on high levels of training, education and investment in research and development as well as networks and partnerships of expertise. The question is whether there can be expertise without experts. While the importance of social sustainability for Arctic expertise has been acknowledged in early Finnish debates, more recently, its social aspects, such as regional innovation potential and creative talent, are not included in the current projects on Arctic expertise. This is in clear contrast to the neighboring countries, Northern Norway and Sweden, and to their understanding of the development of Arctic expertise.

Heidi Sinevaara-Niskanen notes in her chapter that there has been a rather recent shift in the focus in Arctic research and politics toward human development and social sustainability. People of the Arctic have been increasingly recognized as a resource of the region in addition to natural resources. However, as the chapter argues, questions of social sustainability and, in particular, questions of gender have been largely sidetracked from these discussions. This is despite the growing significance that gender plays in Arctic developments, including patterns of migration and demographic changes. Sinevaara-Niskanen sheds light on the connections between gender, equality and sustainability in the Arctic by analyzing social scientific Arctic research and documents produced under the auspices of the Arctic Council. The chapter reveals that, although there is a growing knowledge of gender issues in the Arctic and recognition of the ways in which gender and sustainability mesh, this knowledge has not substantially altered the perceptions of "Arctic resources."

In the following chapter, Francis Joy studies the misuse of religious traditions, cultural heritage and traditional knowledge of the Sámi within the Finnish tourism industry. The appropriation of Sámi traditions is widespread in Finland because the Sámi have not had the resources or power to protect their heritage. Consequently, Sámi shamanism, which is a central practice within traditional Sámi religion, has been adapted in various ways to meet the needs of the tourism industry. Moreover, replica drums and other fake Sámi artifacts, including pretend Gákti costume and other designs, are manufactured and distributed by the tourism industry. The legislation as of 2003, which was meant to protect Sámi cultural heritage, is ambiguous, as are guidelines within the tourism industry. There is no effective legal protection in place against copying and exploiting Sámi cultural heritage in Finland.

In the final chapter of the second part, Susanna Pirnes discusses the role of history in the social imaginaries of the Russian Arctic. Victorious history is often utilized to amplify the chosen political path, which is very true in any Arctic state. In Russia, polar exploration and the Arctic have been an important part of national history and a source for many heroic stories, which were used during the Soviet era for political purposes. Historical knowledge of people is an instrument through which it is possible for the political elites to use certain fragments of history as a culturally sustainable resource. The question of historical knowledge about the Arctic is deliberated with the concepts of historical and culture identity politics.

The concluding chapter by Monica Tennberg, Hanna Lempinen and Susanna Pirnes draws together the main points and lessons from current debates about social and cultural (un)sustainability in the Arctic from the perspective of social imaginaries. Social imaginaries are essential for understanding complex entanglements of resources and sustainabilities, being as they are the very

> ways people imagine their social existence, how they fit together with others, how things go on between them and their fellows, the expectations that are normally met, and the deeper normative notions and images that underlie these expectations.
>
> (Taylor, 2004, p. 23)

References

Boström, M. (2012) "A missing pillar? Challenges in theorizing and practicing social sustainability: an introduction to the special issue," *Sustainability: Science, Practice & Policy*, 8(1), pp. 3–14.

Bridge, G. (2009) "Material worlds: natural resources, resource geography and the material Economy," *Geography Compass*, 3(3), pp. 1217–1244.

Carson, R. (1962) *Silent spring*. New York: Houghton Mifflin.

Del Río, P. and Burguillo, M. (2008) "Assessing the impact of renewable energy deployment on local sustainability: towards a theoretical framework," *Renewable and Sustainable Energy Reviews*, 12, pp. 1325–1344.

Dessein, J., Soini, K., Fairclough, G. and Horlings, L. (eds.) (2015) *Culture in, for and as sustainable development: conclusions from the COST Action IS1007 investigating cultural sustainability*. Jyväskylä: Jyväskylä University Press.

Ferry, E. (2016) "Gold prices as material-social actors: the case of the London gold fix," *The Extractive Industries and Society*, 3, pp. 82–85.

Fondahl, G. and Wilson, G. (eds.) (2017) *Northern sustainabilities: understanding and addressing change in the circumpolar world*. Cham: Springer.

Hiedanpää, J., Jokinen, A. and Jokinen, P. (2012) "Making sense of the social: human–nonhuman constellations and the wicked road to sustainability," *Sustainability: Science, Practice & Policy*, 8(1), pp. 40–49.

International Union for Conservation of the Nature (IUCN). (1980) *World conservation strategy: living resource conservation for sustainable development* [Online]. IUCN–UNEP–WWF. Available at: https://portals.iucn.org/library/efiles/documents/WCS-004.pdf (Accessed: April 18, 2019).

Kassel, K. (2012) "The circle of inclusion: sustainability, CSR and the values that drive them," *Journal of Human Values*, 18(2), pp. 133–146.

Lähde, V. (2015) "Politics in a world of scarcity" in Bergnäs, K., Eskelinen, T., Perkiö, J. and Warlenius, R. (eds.) *The politics of ecosocialism: transforming welfare*. London: Routledge, pp. 55–67.

Latour, B. (2005) *Reassembling the social: an introduction to actornetworktheory*. Oxford: Oxford University Press.

Lempinen, H. (2018) *The elusive social: remapping the soci(et)al in the Arctic energyscape*. Rovaniemi: Lapland University Press.

Meadows, D., Meadows, G., Randers, J. and Behrens III, W. (1972) *The limits to growth*. New York: Universe Books.

Mercer, C. (2002) *Towards cultural citizenship: tools for cultural policy and development*. Stockholm: The Bank of Sweden Tercentenary Foundation & Gidlunds förlag.

Missimer, M., Robèrt, K. Broman, G. and Sverdrup, H. (2010) "Exploring the possibility of a systematic and generic approach to social sustainability," *Journal of Cleaner Production*, 18, pp. 1107–1112.

Murphy, K. (2012) "The social pillar of sustainable development: a literature review and framework for policy analysis," *Sustainability: Science, Practice & Policy*, 8(1), pp. 15–29.

Nilsson, A. and Filimonova, N. (2013) *Russian interests in oil and gas resources in the Barents Sea*. Stockholm Environment Institute Working Paper 2013:5. Stockholm: Stockholm Environment Institute.

Nuttall, M. (2002) "Global interdependence and Arctic voices: capacity-building for sustainable livelihoods," *Polar Record*, 38(206), pp. 194–202.

Nuttall, M. (2010) *Pipeline dreams: people, environment and the Arctic energy frontier*. Copenhagen: IWGIA.

Partridge, E. (2014) "Social sustainability" in Michalos, A. (ed.) *Encyclopedia of quality of life and wellbeing research*. Dordrecht: Springer, pp. 6178–6186.

Petrov, A., BurnSilver, S., Stuart Chapin III, F., Fondahl, G., Graybill, J., Keil, K., Nilsson, A., Riedlsperger, R. and Schweitzer, P. (2017) *Arctic sustainability research: past, present and future*. London: Routledge.

Pram Gad, U., Jakobsen, U. and Strandsbjerg, J. (2017) "Politics of sustainability in the Arctic: a research agenda" in Fondahl, G. and Wilson, G. (eds.) *Northern sustainabilities: understanding and addressing change in the circumpolar world*. Cham: Springer, pp. 13–23.

Pram Gad, U. and Strandsbjerg, J. (eds.) (2019) *The politics of sustainability in the Arctic: reconfiguring identity, time, and space*. London: Routledge.

Psaridikou, K. and Szerszynski, B. (2012) "Growing the social: alternative agrofood networks and social sustainability in the urban ethical foodscape," *Sustainability: Science, Practice and Policy*, 8(1), pp. 30–39.

Raco, M. (2005) "Sustainable development, rolled-out neoliberalism and sustainable communities," *Antipode*, 37(2), pp. 324–347.

Sen, A.K. (1999) *Development as freedom*. New York: Anchor Books.

Slootweeg, R., Vanclay, F. and van Schooten, M. (2001) "Function evaluation as framework for the integration of social and environmental impact assessment," *Impact Assessment and Project Appraisal*, 19(1), pp. 19–28.

Soini, K. and Birkeland, I. (2014) "Exploring the scientific discourse on cultural sustainability," *Geoforum*, 51, pp. 213–223.

Taylor, C. (2004) *Modern social imaginaries*. London: Duke University Press.

Tennberg, M. and Lempinen, H. (2015) "Sosiaalista kestävyyttä etsimässä: tapaustutkimuksena Salla," *Kosmopolis*, 45(3), pp. 27–43.

UN (2015) *UN sustainable development goals 2015–2030*. Available at: www.un.org/sustainabledevelopment/sustainable-development-goals/ (Accessed: April 18, 2019).

Vallance, S., Perkins, H.C. and Dixon, J.E. (2011) "What is social sustainability? A clarification of concepts," *Geoforum*, 42, pp. 342–348.

Vanclay, F. (2002) "Conceptualising social impacts," *Environmental Impact Assessment Review*, 22, pp. 183–211.

Vanclay, F. (2003) "International principles for social impact assessment," *Impact Assessment and Project Appraisal*, 21(1), pp. 5–11.

WCED. (1987) *Report of the World Commission on Environment and Development: our common future*. Oxford: Oxford University Press.

Wilk, R.R. (2002) "Culture and energy consumption" in Bent, R., Baker, R. and Orr, L. (eds.) *Energy: science, policy, and the pursuit of sustainability*. Washington, DC: Island Press, pp. 109–130.

Part I

Entangled resources and sustainabilities

2 Greenland and the elusive better future

The affective merging of resources and independence

Marjo Lindroth

Introduction

In May 2016, during my research visit to Nuuk, Greenland, I followed a demonstration that proceeded along the main shopping street, finally stopping in front of the Inatsisartut, the Greenland parliament. The people were holding up signs with the words "Urani? Naamik" – "Uranium? No." The demonstration was part of a series of many in Nuuk and elsewhere in Greenland since its parliament voted in favor of removing the country's zero tolerance policy on uranium mining in 2013. The vote to lift the ban was extremely narrow, with one vote more for those in favor of starting to mine for uranium and two abstentions. Reflecting this narrow margin in Inatsisartut, the general opinion in Greenland was split into opposite camps as well. The debate over uranium is one among many over resources and their prospective use in Greenland. The debates are fueled by the country's stated political aim of becoming independent from its former colonizer Denmark. The extraction of the potentially vast supply of natural resources is seen as critical for the economic self-sufficiency of the country and for it to ultimately reach independence.

This chapter critically assesses the entanglement of the political pursuit of independence and the plans for resource extraction in Greenland. Resources engender hopes for increased wealth and are key to nation building. The aim of independence induces hopes for a redeemed future, one in which Greenland would be equal among other nations, no longer economically or politically dependent on Denmark. The entwinement of resources and independence also creates different and often contradictory meanings that can stir up fear and concerns over, for example, the environment, culture and the viability of a resource-based economy. Both the creation of a functioning resource extraction industry and the reaching of independence are currently elusive goals in Greenland but ones that are firmly on the agenda and thus leave their mark on the present. Certain political visions for the future, in particular the economic imperative of nation building, have become especially powerful in Greenland.

The chapter offers an investigation of how the hope-laden and mutually inseparable visions for resources and independence engender a complex debate, one that would not be produced by one part of the equation alone. It claims that by drawing on "affective resonances" (Weszkalnys, 2016) that the envisioned developments incite, the debate creates powerful visions for the future of Greenland. In these visions, affective elements that are incompatible with the economic imperative of nation building are often ignored, downplayed or overlooked. My aim is thus not to evaluate whether Greenland will eventually reach full independence or not, nor is the aim to take a stand on if or how Greenland should use and extract its natural resources. Instead, the chapter dissects the ways in which visions for the future of Greenland, elusive as they may be, already have an impact on the country's present and its political, economic and social development.

The chapter builds on discussions of developments in Greenland in terms of resource extraction or the pursuit of independence. In addition, the reflections in the chapter benefit from research interviews that I conducted in Nuuk, Greenland, in May 2016. I conducted three interviews with members of the Greenland parliament and five interviews with civil servants from different ministries (altogether eleven persons). Here, these interviews, referred to as discussions, provide points of reflection on the complexity and contradictory nature of the debate on resources and independence in Greenland. They illustrate that, while there is wide agreement on certain points, such as the necessity of resource extraction, there are also many differences, such as those on the scope and timeline of the envisioned developments for resource extraction.

By bringing in a discussion of how different affective elements feature in the debate on Greenland's future, the chapter makes a contribution to scholarship that has largely focused on issues of the economic and political (un)reachability of independence and resource projects (e.g. Bjørst, 2016; Kuokkanen, 2017; Wilson, 2017), democracy and (lack of) public participation in resource developments (e.g. Wilson, 2016; Pelaudeix et al., 2017) or the more technical questions of the "what, where, how" of resource extraction (Boersma and Foley, 2014). Earlier research has acknowledged the connections between plans for resource extraction and the pursuit of independence, but their entwinement is more complex than has been recognized. As a notable exception in the existing scholarship, Mark Nuttall (2012, 2013, 2017) has, in his extensive research on Greenland and resource politics, discussed how the country's future is debated through hopes and aspirations. My chapter continues this line of scholarship by further elaborating the affective elements in the debate.

Independence: unrealistic yet inevitable?

For the past ten years, Greenland has had the legal right to become independent from Denmark, and independence has become a central question in the discussions of the nation's future. Greenland has gradually gained more autonomy from Denmark, with the developments culminating in 2009 in the Act on Greenland Self-Government coming into force. According to the Act, the people of Greenland are recognized as a separate people under international law; thus, the Act recognizes that the people of Greenland have the right to become independent if they so wish. The Act also states that any revenue from the country's natural resources belongs to Greenland. For Greenland, as a nation striving for independence from its (former) colonizer, gaining increased economic self-sufficiency has become the dominant political aim envisioned to enable the reaching of independence.

The pursuit of independence in Greenland deals with an issue that has mattered to all nations that have been under colonial rule: it is about reaching equal status among other nations and about (re)gaining control of the future development of the country. The politically stated aim of independence draws on the hopes and aspirations for freedom, self-determination, pride, self-worth and equality that are ingrained in visions of being an independent nation. Along the same lines, and echoing the view of independence as an inevitability and an almost predestined course of events, similar sentiments were voiced in my discussions in Nuuk. Independence was talked about as something that is "healthy" and the pursuit of Greenland's independence as something that is "natural" and a logical development in history (Nuuk, May 12, 2016).

It is clear that the current pursuit of Greenland's independence is intrinsically bound to the resource potential of the country. In political parlance, then, independence is envisioned as an inevitable development and an unstoppable outcome once Greenland manages to establish a well-functioning industry for the extraction of its natural resources (Bjørst, 2016; Wilson, 2017). Overall, my discussions in Nuuk revealed that there is strong unanimity that resources will facilitate the pursuit of independence in Greenland (Nuuk, May 11, 2016). However, there is wide variation among the Greenlandic political parties as to which of the country's resources should be extracted, and how, to what extent and when they should be used. The case on uranium mining, presented at the beginning of the chapter, is a vivid example of the differences and contentious issues in the debate.

While the commonly held political ambition in Greenland is to reach independence, there are varying opinions on this matter as well in terms of, for example, on the way forward, on the time frame for the process and on how realistic the actualization of the goal is. My discussions in Nuuk illustrated this as there were views emphasizing that Greenland has all the independence it needs, while according to another perspective, all ties with Denmark should be severed as soon as possible. According to the latter view, independence is the only answer to the social ills of Greenlandic society, ills that are caused by the former colonizer Denmark (Nuuk, May 11, 2016). Most views are likely to be somewhere in between these opposites. There were calls for pragmatism and realism, emphasizing the need to first address social issues while holding the goal of political and economic independence firmly in sight (Nuuk, May 24, 2016).

Regardless of how independence is prioritized, it is a question that cannot be ignored or bypassed in Greenlandic politics. Independence was perceived, in some of my discussions in Nuuk, as something that is politically correct to be on the agenda in Greenland, but that it is a dream rather than an actual goal. In this light, independence was seen as unrealistic because Greenland would not really be independent as other such small nations cannot be (Nuuk, May 11, 2016). Independence was more akin to a utilizable dream that politicians employ in order to gain votes (Nuuk, May 17, 2016). The pursuit of independence was also seen as a rhetorical and populistic move that plays on dichotomies that is not realistic geopolitically, demographically or economically (Nuuk, May 12, 2016).

The discussion of the future independence of Greenland is, on the one hand, about questions of what is realistic to achieve for such a small nation and, on the other hand, about views that see independence as an unstoppable end point of development. In the latter vision, the end point refers both to economic self-sufficiency as the outcome of a functioning resource extraction industry and to full political independence as the result of decolonization and nation building. However distant the actualization of independence might be, it becomes politically constructed as an inevitability based as it is on the vast and varied resources that Greenland already possesses.

(Resource) developments of contradiction

As the government of Greenland's oil and mineral strategy for the years 2014 through 2018 illustrates, resources are a source of hopes and positive expectations:

> So long as the mineral resources remain in the subsoil, they will create no value for Greenland. Conversely, active mining and/or oil/gas extraction activities will bring more jobs, more contracts and more revenues for the treasury.
>
> (Government of Greenland, 2014, p. 19)

Greenland's resources, in terms of those areas of the country that have been explored, are rich and varied. There are minerals, potential for offshore oil and iron ore, copper, zinc, gold, uranium and rare earth elements. According to many accounts, the promise is that as larger areas of the country are explored, more significant resource findings will be made (Boersma and Foley, 2014). As the oil and mineral strategy quoted here demonstrates, the subsoil of the country is presented as the solution for a challenging economic and social situation. In political visions for the future of Greenland, the subsoil is narrated into being as a place of promise, wealth and potentiality, creating jobs where people are in dire need of employment and where the well-being of communities is on the line (see also Nuttall, 2017). This political will to extract the resources is justified on the basis of the economic and social needs of the Greenlandic people themselves.

Tapping into Greenland's natural resources is politically legitimized not only on the basis of the envisioned increase in the wealth of the nation and the well-being of the population but also because it is deemed crucial for reaching the aim of independence. Resources engender hopes for profit and thus hold the key to Greenland's economic and political self-sufficiency. This is despite the challenges and drawbacks that the unpredictability of global markets creates, as Dodds and Nuttall (2018, p. 151) have observed:

> Despite a dip in global commodity prices, as well as other global processes, the subterranean and the ocean depths nonetheless remain critical for Greenlandic notions of nation-building and state formation.

Currently, the country's economy is highly dependent on fishing and on the annual block grant it continues to receive from Denmark. The block grant forms a large part of the country's budget revenue (Nuttall, 2012). The Act on Greenland Self-Government (2009) gives Greenland the authority over its natural resources, which includes deciding the way forward on if and how to start extracting and using them. The annual block grant from Denmark, according to the Act, will be gradually reduced when Greenland begins gaining revenue from the extraction of its natural resources. This is crucial, as it makes the stakes very high for Greenland. In order to be able to sustain its economy and build an independent future, the country needs the necessary political, legal and economic frameworks and the appropriate infrastructure in order to attract investment and resource extraction projects, as is also stated in the oil and mineral strategy:

> In the current economic climate, achieving the objective of increasing self-sufficiency is not realistic without substantially developing the mineral resources sector.
>
> (Government of Greenland, 2014, p. 17)

Resources are also sources of worry and anxiety that predict destruction, risk and threat (Bjørst, 2017, p. 30). Plans for resource extraction projects often engender fears that the nature, land and sea will be negatively affected through pollution or changes in and sometimes even destruction of landscapes. Indeed, it is a cause for concern that in the name of pursuing state formation and independence, there will be a price to pay in terms of the environment. The envisioned projects also raise concerns that traditional livelihoods and ways of using the land and the sea will be severely impacted and threatened.

While most would agree that resource extraction is needed in order for the country to provide for itself in the future, there are views that caution how the promise of wealth can often distract from what takes place on the ground and in the everyday. This worry was also echoed in my discussions in Nuuk, where the values that Greenland is basing its independence on were questioned. The sustainability of giving mines the role of the saviors of the economy was similarly challenged because mines will be there for a while and then won't (Nuuk, May 24, 2016). Indeed, the lifespan of the projects is a source of worry, and a common concern is what happens to the viability of the communities after the resources are exhausted. The visions for resource extraction also engender doubts about whether, after they are up and running, the projects will actually benefit and profit communities in Greenland or whether the promised revenue will, in fact, flow outside of the country (Nuttall, 2013; Brincker, 2017).

The power that is entailed in the pursuit of independence through resource extraction becomes manifest in the silences it creates. For example, other ways of perceiving and imagining the landscape – different or alternative to resource extraction – are often not seen or properly recognized. Mark Nuttall (2017) has described how certain areas become constructed as resource spaces by business and political elites in order to attract investments, resource exploration and exploitation, through various activities of mapping etc. These resource spaces, and those parts of Greenland that are being built as such, risk becoming "zones of sacrifice" (Nuttall, 2017, p. ix). These zones are vital for developing mining, gaining revenue and thus increasing powers of economic autonomy. At the same time, however, they unmake places imbued with human history and experience, rendering human presence in the landscape invisible and thus inconsequential (Nuttall, 2017, p. ix). In these scenarios, the views and interests of the local population can often be very different than the interests of the government or business elites (Wilson, 2016, p. 74).

So far, the government of Greenland has issued mainly exploration licenses while only a few extraction licenses have been issued (Brincker, 2017). Despite all the promising visions, the economic potential of resource extraction projects and their materialization are still largely at the level of estimations and more or less vague possibilities, as also becomes evident in Greenland's oil and mineral strategy:

> Our best guess at the moment is that three to five mines may be opened within the next five years, and it is estimated that one to two offshore drilling projects will be established every second year.
> (Government of Greenland, 2014, p. 15)

However, it is crucial to note how the mere anticipation of resource extraction projects can start to powerfully determine, define and guide what is thought of as desirable, doable and sensible in the present (Sejersen, 2015). Greenland, as the new Arctic resource frontier, invites entrance by foreign companies and investment as its resources are already there waiting to be extracted. In these visions, the resources will inevitably be explored and exploited, based on their huge promissory profit, which is still largely unrealized. These figurations are often presented as cold and rational economic facts, devoid of politics. This is a move that enables current politics to transcend the given state of affairs, a "venture beyond" (Meyer, 2010, p. 110) that gives the badge of rationality, "there was no alternative" and even patriotism to political (in)actions that will allegedly shape the future in favor of the Greenlanders.

In sum, the repercussions of the quest for economic self-reliance do not stay within the realm of economics. Although the resources are not (yet) used or extracted, different visions of them already leave a strong mark on social and political life in Greenland (see also Guerrieri, 2019a, on visions for oil and gas developments in Northern Canada). In this debate about Greenland's potential, resources are given different meanings by different stakeholders. These debates are not just political but ideological as well (Nuttall, 2017, p. ix). The economic imperative in Greenlandic politics reinforces the view that resource extraction is an inevitability and renders other concerns – over the environment, security, culture – less pressing.

The better future that keeps escaping

Both visions for Greenland, on independence and on the use of natural resources, hinge on hopes for a better future. The urgency to realize the resource potential of the country – in order to mend unemployment and other social ills and to meet the rising costs of an aging and out-migrating population – further fosters the view of resources as the saving grace of Greenland. The positive tone of the political discourse on resource extraction sees it as a force for change in a situation where, for many in Greenland, staying the same is the worst scenario of all (Bjørst, 2016, p. 37).

Resource extraction brings hope for a better and more prosperous future as it entails the promise of material and societal improvement. In this rhetoric, resource extraction is necessary and something that is unavoidable, envisioned as it is to create profit and well-being for Greenlanders and, ultimately, an independent future (Nuttall, 2008, p. 65; Wilson, 2017, p. 512). In this vision of the future, Greenland cannot not mine. As Bjørst notes (2016, p. 35), the political rationale of this hopeful vision is that mining would be done *for* Greenland, not just *in* Greenland. It seems a fact that large resource extraction projects will eventually become a reality, but the time frame for this remains unknown. Currently, there are many more exploration licenses granted, and actual extraction activities are very few. In order for Greenland's natural resources to sustain and develop the country's economy, the projects need to materialize, but to date this development has been very slow.

A similar elusiveness haunts the pursuit of independence in Greenland. The legal right to become independent, if the people of Greenland so wish, looms large in the public and political visions for the future of Greenland. However, there is a continuous fluctuation in how the reachability of independence is perceived. The reaching of this "end point" keeps escaping with, for example, a decrease in global demand for minerals and resources, only to appear closer again with the finding of new potential resource sites and a rise in global demand and market prices. As Weszkalnys (2016, p. 142) points out in her study of the societal effects of different resource visions, anticipation, worries and trepidations "intensify, fade, and circulate as resource prospects are predicted, pursued, and abandoned," and, in Greenland, along with this fluctuation, the hopes and fears connected to independence also ebb and flow.

The expected impacts of climate change contribute to this conundrum of visions for the future in Greenland. While climate change poses a threat to the traditional livelihoods and practices of the Greenlandic Inuit and the country's environment, it also creates hopefulness by (possibly) opening up new possibilities to start exploring and exploiting the country's resources and by making them more accessible. As Mark Nuttall (2017) has noted in connection with those involved in the Greenlandic political and business sector, many of them see climate change, rather than a threat to the environment, "as something that will allow the implementation of empowering political and economic decisions" (Nuttall, 2017, p. 27). Indeed, while climate change poses threats to Greenlandic culture and livelihoods, thus echoing the global image of climate change as a "looming catastrophe," it can also be viewed as providing new opportunities for sustaining lives, livelihoods and cultures – and enabling state formation (Nuttall, 2012, p. 118). Climate change can provide Greenland with increased access to its resources and possibly enhance interest and investment from elsewhere, for example China or the EU (Pram Gad, 2017, p. 73), and thus facilitate achieving full independence from Denmark.

Both of these hard-to-reach yet potentially attainable objects in Greenland – resource extraction and independence – are "already there": the natural resources in the subsoil and under the sea and the right to independence written down in the Greenland Act of Self-Government. This is the base for "speculation, hope and anticipation" in the discussions of the future of Greenland (Nuttall, 2017, p. x). Greenland is in a simultaneous state of potentiality and elusiveness: a situation where there is anticipation of developments in terms of resources, more exploration and projects and thus more self-sufficiency, but these may and will take a while. It is an "in-between-ness" where every now and then expectations are fostered, until development slows down once more.

Even though the actualization of resource extraction projects and independence keeps escaping into the future, a position of hope-laden anticipation is created. The resources of Greenland have a promissory role, one that encourages confidence in the inevitability of future social and economic development and independence. In the words of Valeria Guerrieri (2019b, p. 690, on resource developments in the Canadian Northwest), "an inescapable site of promise" is carved out, one which "traps" those involved through giving them hope and becoming the common-sense way of seeing the situation. In Greenland, this dynamic draws on the real need to improve the current social and economic situation of the country that urgently calls for hope and brighter future scenarios.

Rational vs. irrational

The actualization of Greenland's independence has become conditional on the country being successful in developing its resource extraction industry. With this inseparable entwinement, there is a lot at stake in Greenland in terms of political, economic, cultural and social questions and environmental impacts. The development of resource extraction, and especially the prospects for uranium mining, have become both a threat to the very existence of Greenland and a necessity that will secure the country's future (Søby Kristensen and Rahbek-Clemmensen, 2018, p. 39). Thus, the debate is loaded with various and often contradictory meanings, possibilities and threats. Hopefulness and anticipation are accompanied by other, less optimistic states of doubt, disillusionment and even fear (see also Weszkalnys, 2016, p. 128, on resource affects).

Being doubtful of, opposed to or critical of resource extraction can gain meanings that go far beyond or do not have much to do with the development of particular projects. The debate about the future in Greenland is fueled by past (and current) experiences of disparity and inequality in relation to Denmark. As Kirsten Thisted (2017, p. 233) points out, resource extraction has become the vehicle through which this past is envisioned to be replaced by pride and equal relations between the two nations. The question is no longer only about the viability of a particular community, but ultimately about the future prosperity and independence of the whole nation. This also came up in the discussions I had in Nuuk. One of the themes was how the critical importance of independence – and the resources that are seen to enable it – often make the political pursuit of independence a backlash against the colonial past, one that can go overboard (Nuuk, May 11, 2016). Ulrik Pram Gad (2017, p. 112) has also noted how the distorted power relations between Denmark and Greenland and the colonial past – often involving maternalism and the infantilization of Greenland by Denmark – ignite a strong counterreaction in Greenland. In these debates, Greenland's independence easily becomes, in the words of Søby Kristensen and Rahbek-Clemmensen (2018, p. 48), "a cherished object that is threatened."

The perception that Greenland's future independence is being threatened can lay the ground for a politics that takes place at the cost of public participation, openness and transparency. My discussions in Nuuk also referred to this problem as there were views that the government of Greenland at that time had failed to respect democracy (Nuuk, May 24, 2016). In a situation where the country has a pressing need for resources, income and foreign investment, it becomes pertinent to ask, as Emma Wilson has noted (2016, p. 74), whether the government can "grant consent on behalf of local communities." Indeed, many scholars have observed how local understandings and ways of using the land risk being bypassed as a result (e.g. Nuttall, 2013; Kuokkanen, 2017; Pelaudeix et al., 2017; Søby Kristensen and Rahbek-Clemmensen, 2018). For example, the particularly controversial issue of uranium mining and exporting has fired demands for a referendum on the matter but to no avail. This stirs up fears in the opponents of uranium mining that they will not be heard (Price Persson and Sommer Christiansen, 2016; McGwin, 2018).

It is noteworthy that the less heartening aspects of the debate are not only about whether or not the local people are heard or whether or not the environmental and societal impacts of resource projects are properly assessed. They are about ignoring or overlooking certain affective elements that the entanglement of resource extraction developments and the pursuit of independence invokes (on affective resonances and resources, see Weszkalnys, 2016, p. 127). Economic facts with the plans for resource extraction and their promissory, so far unrealized profits tend to make the economically-centered arguments seem rational and the only viable choice. Accordingly, anything that goes against this narrative is easy to call out as irrational and even immature (Søby Kristensen and Rahbek-Clemmensen, 2018, p. 45). Frank Sejersen (2015, p. 119) has noted how opposition to resource projects is often labeled unpatriotic in Greenland as it is seen as something that will block national development. Opposing the economic rationality thus becomes subject to moral judgment and, as a "disturbing" position, also an object of political action (Weszkalnys, 2016, pp. 128, 133; Stoler, 2004; see also Dawney, 2018, on how power works through affective states). The economic imperative of nation building in Greenland relies on the promise of independence, national pride, revenue and employment and on labeling other, more suspicious or concerned voices as hindering the project. It calls for certain sacrifices to be made today – for example, in terms of the environment, cultural values or democracy – in order to (supposedly) redeem its promise in the future of being an independent nation.

To be sure, resource development often raises controversial views in the local communities, and Greenland is no exception to this as a country striving to come to terms with the relationship between its resource potential and its cultural and social values. There only has to be a mere anticipation of resource findings and potential extraction to engender struggles over the ethically, culturally or environmentally appropriate directions for the future in terms of resource extraction (Weszkalnys, 2016, p. 129). In Greenland, the pursuit of independence, which is the cherished object under threat, further revs up this debate as it makes the stakes for Greenland even higher.

Conclusions

In a political, economic and social sense, maintaining the status quo is not an option in Greenland. The past experiences of inequality and colonialism, in addition to the current social and economic situation, characterized as it is by an urgent need for improvements, call for hope and brighter future scenarios. The vast resource potential of the country has been politically harnessed as the answer that will, with time, provide economic self-sufficiency and independence.

While the goal of independence and the development of resource extraction are very much on the political agenda, they still remain elusive. This makes the debate especially powerful as precious political goals, such as independence, can often be viewed as being under threat. Despite its elusiveness, the potential better future has a strong presence in contemporary politics in Greenland. Envisioned as it is to create the much-needed revenue and ultimately enable full independence, the development of the resource potential of Greenland seems an unavoidable fact, albeit one that may take some time to actualize.

The political pursuit of independence hinges on resources and, as a result, it is a powerful engine of development, so much so that it often trumps other ways of envisioning the future. The entanglement of Greenland's independence with the country's resources creates binaries that predict salvation/destruction, promise/threat, hope/fear. They "organize" and construct the debate as a package that is understandable and manageable in a certain way. In particular, the promissory nature of resources and independence forms the basis for speculative visions of increased wealth and self-sufficiency. The economic imperative of national development, one to which there seems to be no alternative, easily becomes the "proper" or "rational" way of envisioning the future and of making political decisions today. Furthermore, the past experiences of colonization are very much present in how resource extraction is perceived and envisioned in Greenland and the meanings that are attached to resources. They have become the vehicle through which it (allegedly) becomes possible to relegate the colonial history firmly to the past. Resources are the key that provide Greenland with the opportunity to reach an equal status among other nations.

As political goals, independence and resource extraction guide the development of the country. However, the debate on Greenland's future brings forth the more complex ways in which power is entailed in the entanglement of resource extraction and the pursuit of independence. To dissect this complexity, it becomes pertinent to examine the role that affective elements have for contemporary politics in facilitating or preventing certain types of developments. Currently, it seems that the potential threats connected to resource extraction yield to the hope and political promise that resources are envisioned to carry for the future of Greenland. These hope-laden resource visions thus function as the more powerful engine of the country's political and economic development.

References

Act on Greenland Self-Government (2009) [Online]. Available at: https://naalakkersuisut.gl/~/media/Nanoq/Files/Attached%20Files/Engelske-tekster/Act%20on%20Greenland.pdf (Accessed: February 14, 2019).

Bjørst, L.R. (2016) "Saving or destroying the local community? Conflicting spatial storylines in the Greenlandic debate on uranium," *The Extractive Industries and Society*, 3(1), pp. 34–40.

Bjørst, L.R. (2017) "Uranium: the road to 'economic self-sustainability for Greenland'? Changing uranium-positions in Greenlandic politics" in Fondahl, G. and Wilson, G. (eds.) *Northern sustainabilities: understanding and addressing change in the circumpolar world*. Cham: Springer, pp. 25–34.

Boersma, T. and Foley, K. (2014) *The Greenland gold rush: promise and pitfalls of Greenland's energy and mineral resources* [Online]. Energy Security Initiative at Brookings and John L. Thornton China Centre at Brookings. Available at: www.brookings.edu/wp-content/uploads/2016/06/24-greenland-energy-mineral-resources-boersma-foley-pdf-2.pd (Accessed: August 27, 2018).

Brincker, B. (2017) "Images of the Arctic: visualising Greenland as an indigenous people and a modern nation," *Visual Studies*, 32(3), pp. 251–261.

Dawney, L. (2018) "The affective life of power," *Dialogues in Human Geography*, 8(2), pp. 218–219.

Dodds, K. and Nuttall, M. (2018) "Materialising Greenland within a critical Arctic geopolitics" in Søby Kristensen, K. and Rahbek-Clemmensen, J. (eds.) *Greenland and the international politics of a changing Arctic. Postcolonial paradiplomacy between high and low politics*. Abingdon: Routledge, pp. 139–154.

Government of Greenland (2014) *Greenland's oil and mineral strategy 2014–2018* [Online]. FM 2014/133. Available at: https://naalakkersuisut.gl/~/media/Nanoq/Files/Publications/Raastof/ENG/Greenland%20oil%20and%20mineral%20strategy%202014-2018_ENG.pdf (Accessed: October 10, 2018).

Guerrieri, V. (2019a) *Not there and (not) yet. The spatio-temporal politics of oil and gas development in the Canadian North*. PhD thesis, Faculty of Humanities, University of Copenhagen.

Guerrieri, V. (2019b) "The spatiality of hope: mapping Canada's Northwest energy frontier," *Globalizations*, 16(5), pp. 678–694.

Kuokkanen, R. (2017) " 'To see what state we are in': first years of the Greenland Self-Government Act and the pursuit of Inuit sovereignty," *Ethnopolitics*, 16(2), pp. 179–195.

McGwin, K. (2018) "As Greenland nears uranium decision, opponents fear public won't be heard," *Arctic Today*, May 16 [Online]. Available at: www.arctictoday.com/greenland-nears-uranium-decision-opponents-fear-public-wont-heard/ (Accessed: October 25, 2018).

Meyer, I. (2010) "Hope as the conscious action towards an open future" in Horrigan, J. and Wiltse, E. (eds.) *Hope against hope. Philosophies, cultures and politics of possibility and doubt.* Amsterdam: Rodopi, pp. 97–111.

Nuttall, M. (2008) "Self-rule in Greenland: towards the world's first independent Inuit state?" *Indigenous Affairs*, 3–4, pp. 65–70.

Nuttall, M. (2012) "Imagining and governing the Greenlandic resource frontier," *The Polar Journal*, 2(1), pp. 113–124.

Nuttall, M. (2013) "Zero-tolerance, uranium and Greenland's mining future," *The Polar Journal*, 3(2), pp. 368–383.

Nuttall, M. (2017) *Climate, society and subsurface politics in Greenland. Under the great ice.* London: Routledge.

Pelaudeix, C., Basse, E.M. and Loukacheva, N. (2017) "Openness, transparency and public participation in the governance of uranium mining in Greenland: a legal and political track record," *Polar Record*, 53(6), pp. 603–616.

Pram Gad, U. (2017) *National identity politics and postcolonial sovereignty games: Greenland, Denmark, and the European Union.* Copenhagen: Museum Tusculanum Press.

Price Persson, C. and Sommer Christiansen, M. (2016) "Greenland is divided over uranium mining," *ScienceNordic*, May 27 [Online]. Available at: http://sciencenordic.com/greenland-divided-over-uranium-mining (Accessed: October 17, 2018).

Sejersen, F. (2015) *Rethinking Greenland and the Arctic in the era of climate change: new Northern horizons.* London: Routledge.

Søby Kristensen, K. and Rahbek-Clemmensen, J. (2018) "Greenlandic sovereignty in practice: uranium, independence, and foreign relations in Greenland between three logics of security" in Søby Kristensen, K. and Rahbek-Clemmensen, J. (eds.) *Greenland and the international politics of a changing Arctic. Postcolonial paradiplomacy between high and low politics.* Abingdon: Routledge, pp. 38–53.

Stoler, A.L. (2004) "Affective states" in Nugent, D. and Vincent, J. (eds.) A *companion to the anthropology of politics.* Malden: Blackwell, pp. 4–20.

Thisted, K. (2017) "The Greenlandic reconciliation commission: ethnonationalism, Arctic resources, and post-colonial identity" in Körber, L-A., MacKenzie, S. and Westerståhl Stenport, A. (eds.) *Arctic environmental modernities: from the age of Polar exploration to the era of the Anthropocene.* Cham: Palgrave Macmillan, pp. 231–246.

Weszkalnys, G. (2016) "A doubtful hope: resource affect in a future oil economy," *Journal of the Royal Anthropological Institute*, 22(S1), pp. 127–146.

Wilson, E. (2016) "Negotiating uncertainty: corporate responsibility and Greenland's energy future," *Energy Research & Social Science*, 16, pp. 69–77.

Wilson, P. (2017) "An Arctic 'cold rush'? Understanding Greenland's (in)dependence question," *Polar Record*, 53(5), pp. 512–519.

3   Promise and threat

Living with nuclear in the Finnish context

Hannah Strauss-Mazzullo

Introduction

"If we have a nuclear accident in Finland, it will affect us in any case" was the answer when I asked an informant how they would deal with the risk of living only some ten kilometers away from Finland's next nuclear power plant. He said this with a smile and shrugged his shoulders lightly. Much was left unsaid, as tends to happen when we talk about risk, especially when we take the risk somewhat voluntarily, on our own accord. The threat of dying from exposure to nuclear radiation or physical decay following an accident is only indirectly part of our conversation. Mostly, the theme remains untouched. Using humorous expressions can ease tensions and overcome the problem of talking about the greatest threats. The emotional work behind this expression does not imply that the risk is taken lightly but is an expression of how anxiety is being dealt with in the face of undeniable and continuous threat.

My informant's expression also changed my position from observer/outsider to one affected by the development. And he created a superior position for himself (see Parkhill et al., 2011, p. 332 on empowerment) as he was showing that he was well aware of the general threat we are all living with, whereas I was not considering myself at risk, just like anyone else living at a certain minimum distance from a nuclear power site. According to my informant, the risk of being heavily affected by a radioactive cloud moving to where I, the author, live and where my informants live does not exclude me from those concerned by the development. The cognitive dissonance I experienced following this comment was slowly reduced on the way back home, during a five-hour-long bus journey. The longer the trip took, the more distant the community seemed from where I live. Hence, the 250-kilometer distance makes me believe (foolishly?) that I have a chance to survive the worst-case scenario – a belief that has been challenged by those living close to nuclear facilities: for instance, as cited in Parkhill et al. (2011, p. 333):

> [I]f one of them [reactors] did blow we'd be the first to go so hey, wouldn't necessarily bother us. It would be the people in the middle of the country that would suffer most. We'd be gone so it wouldn't really matter if it really went.
>
> (Maxine, Bradwell)

On the bus, I overheard conversations about the recently announced developer's plan to move its headquarters from Helsinki to this small northern municipality that had agreed to become the third site of nuclear power production in Finland. People on the bus seemed excited, if not proud, but also nervous knowing that so much attention was being paid to their community. Pyhäjoki has suddenly become a place on the map for nuclear power managers as well as anti-nuclear protesters worldwide. Its residents are careful in conversations with outsiders but also among themselves. Meeting an acquaintance by chance at the supermarket or on the bus will not provoke talk about the risks associated with a nuclear power plant. The nuclear power company's name nevertheless dominates much of the conversation on the street as it has contributed large sums to constructing a new, several-story-high building in the city center that will house the town library and an exhibition by the nuclear power company about the ongoing development as well as apartments. The single-story grocery store that will be replaced by the new, multifunctional building is being moved to a field nearby and significantly enlarged to meet growing customer demand. By financially supporting the community, the nuclear power company is working closely with the needs of the current residents. Among the protesters against the development, individuals from Pyhäjoki are rarely seen; more are from surrounding villages, where the benefits are not so clear, and even more are individuals and groups from Helsinki and neighboring Sweden. The low success of the Pro Hanhikivi ("for Hanhikivi," the future site of development) protest groups to draw people to participate in their activities, ranging from information events to artistic workshops to demonstrations, has surprised activists coming together in Pyhäjoki. Moreover, activists have said that their actions have been met with almost violent opposition at times, creating a feeling of impermeable support for nuclear power in the highly cohesive community. The community of Pyhäjoki has a high share of inhabitants who are members of the Laestadian Lutheran Church. The followers of Laestadian belief stand out from the rest of the Finnish population because of high fertility rates, and they tend to keep close company due to their traditional customs. This might be another reason why the community has voiced strong interest in hosting a megaproject with the prospect of employment and continuous revenue (Joona, 2018).

This chapter seeks to understand the societal context and the history of the development in Pyhäjoki and other communities affected by nuclear power or uranium mining projects. The starting point for this research was that both the opposition movement and the nuclear power company were surprised to find certain municipalities competing to host Finland's next nuclear power plant. In combination with governmental support of nuclear power production (Haukkala, 2018) and potentially also uranium mining, and also legislative amendments that simplify extraction and strengthen the nuclear safety watchdog, Finland appears to possess the qualities nuclear managers abroad envy. While most Western nations mull over a nuclear phaseout, especially in the aftermath of the Fukushima incident, Finland is expanding on this resource, despite concerns over the unknown consequences of extracting the fuel, producing electricity and storing the highly radioactive substances after they have been depleted for electricity production. This chapter will thus aim at the bigger picture as well as keeping the local viewpoint in sight. It will further locate the acceptance of high-risk technology theoretically and propose an interpretation of the Finnish case within notions of contentment, trust and an outstanding record of handling nuclear power.

Research for this chapter has been conducted since 2007, focusing on the sociological analysis of nuclear power decision-making and acceptance in the respective Finnish communities, the professionalization of environmental impact assessment procedures and national debates around recent mining developments. I have interviewed local residents, politicians, company managers, environmental activists, environmental authorities and consultants in the communities of Pyhäjoki, Simo, Eurajoki, Simo, Kolari, Oulu and Rovaniemi and have complemented these by observations of public hearings, an examination of official documents in hearing procedures and a review of newspaper articles (mainly *Helsingin Sanomat*, *Kaleva*, *Lapin Kansa*) and by an analysis of online presentations of selected organizations and institutions (e.g. environmental movement organizations, company websites).

Finland's nuclear history and future concerns

Finland is often called a country of "few resources," which emphasizes the lack of fossil fuels in our carbon-intensive society. The mix of energy resources consists of wood, oil, nuclear, coal, gas, hydropower, peat and wind, in this order of importance. Hence, over half of the resources have to be imported, mostly from Russia, while wood, hydropower, peat and wind energy are produced domestically. Of all EU countries, Finland has the highest per capita consumption of energy, with industry being a major consumer of electricity (Official Statistics of Finland [OSF], 2017). Particularly the forestry sector with its pulp and paper production but also the heavy metal industry and the construction sector are among the top consumers. Current energy policy aims to reduce dependency on imports, expand renewable energy production and, at the same time, maintain cheap supply for industrial consumers. While wind power receives the highest public approval in the renewable sector, resistance against wind farms and skepticism toward the integration of unsteady supply into the electricity market continue to slow development. Nuclear power thus plays a crucial role in the current structure and ideological commitment of energy producers and consumers in Finland. Not only does it supply high and steady amounts of electricity, but it also has low carbon emissions and provides the security of experts' being familiar with its handling. It is almost perceived as a domestic resource, despite the fact that fuel currently has to be procured from abroad.

Nuclear power production in Finland is embedded in an international and historical context, so the local development in Pyhäjoki described earlier takes us far beyond national borders and back in time. To date, Finland has two nuclear power plants with two units each (see Map 3.1). They are renowned for their continuous upgrades and uninterrupted production of electricity (capacity load around 95 percent), which means that there are comparatively few times during the course

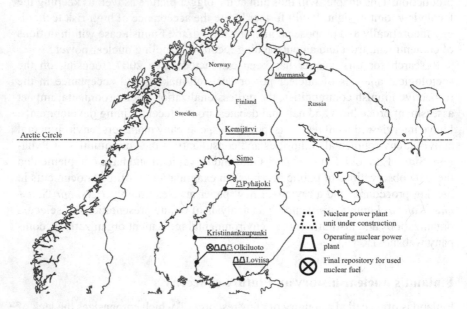

Map 3.1 Actual and future nuclear power production sites in Finland as well as selected
towns included in the assessment process

of the year when the production has to be shut down temporarily (for instance,
to replace the fuel rods). In the 1970s and 1980s, two of the four operating units
were built with Swedish technology and two with Soviet technology in a so-called
"package deal" to appease the Soviet government. Procurement from US suppliers
was not considered as it would have conflicted with the Agreement of Friendship,
Cooperation and Mutual Assistance between Finland and the Soviet Union since
1948 (Müller et al., 2017). The nuclear power plant unit now in the final stages of
construction features French technology. This is important to understanding the
ongoing discussions of Russian involvement in the Fennovoima company, which
is now preparing to build a nuclear power plant in Pyhäjoki (Aalto et al., 2017).
The Russian state is indirectly involved through debt financing of the project, and
the Russian company Atomstroyexport will provide the pressurized water reactor.
This was negotiated in 2016 but not without debate in the Finnish media.

The decision whether or not to use Russian technology stands in the context
of other diplomatic pressures: for instance, Finland's consideration of becoming a member of NATO. The most recent response by the Russian government
was that there would be an immediate (negative) answer of some sort if Finland
took this step (YLE News, 2018). Hence, the installation of nuclear technology
implies some maneuvering between concerns for national security and peaceful relations with the former ruler and global superpower. More recently, major
changes in the company structure of Fennovoima have provoked lively debates

in the Finnish media, especially regarding the involvement of the Russian nuclear power company Rosatom, which currently owns one third of the shares.

The nuclear power company Fennovoima was founded to build another nuclear power plant in Finland. Initially, it consisted of major Finnish industrial companies, which would buy electricity from Fennovoima on cost basis, i.e. no additional profit would be made by Fennovoima. This is called the Mankala principle in accordance with the name of the company that pushed for this model in the 1960s. Most Finnish nuclear power companies and many other businesses currently work according to this model, which secures cheap electricity for energy-intensive industry over decades. In 2010, the model was officially contested by Green politicians Satu Hassi and Heidi Hautala in a written request to the European Commission (E2217–10) to investigate unfair competition. The case was closed in 2012 without further action.

In Finland, the potential host municipality has a veto right on the site of nuclear development. Only if decisions are positive at the municipal and governmental level ("decision in principle") can the developer submit an application for a construction permit. During construction, the Radiation and Nuclear Safety Authority (STUK) watches over the fulfillment of procedural, legal and technical requirements. The safety authority is the most trusted institution in the Finnish energy sector (Choi, 2018) and has communicated information to citizens efficiently and mostly satisfactorily (Kari et al., 2010). In order to improve communication, STUK went as far as providing training to journalists on nuclear issues and has thus become the first source contacted for any questions. At the same time, Litmanen et al. (2017, p. 26) note that "local activists have complained that it was hard to find independent expertise in a small country like Finland where almost all the experts were somehow connected either to the industry or the regulatory authorities." STUK's competences have been extended in a recent regulatory amendment of the Nuclear Energy Act in 2015. In the "strong administrative state" of Finland (Säynässalo, 2009), elite and technocratic decision-making continues to determine siting procedures of nuclear facilities.

> Both in Finland and Sweden regulatory regimes resemble fairly conventional state-centric arrangements, where dyadic relationships between the industry and the regulator are the norm, leaving almost no room for civil society actors.
>
> (Litmanen et al., 2017, p. 21)

In general, the Finnish handling of nuclear technology, be it of Soviet, Swedish or French origin, is highly esteemed among Finnish citizens. Worries have mostly surfaced over the employment of foreign nationals at the construction site of Olkiluoto 3, Finland's soon-to-be fifth nuclear power plant unit in the municipality of Eurajoki. There, up to 1,000 Polish workers have been employed by the initially French-German consortium, together with Portuguese foremen and French technicians. The nuclear power plant unit Olkiluoto 3 is a multinational project that has struggled with massive cost overruns and significant delays.

The political and managerial context of building a nuclear power plant in the 21st century entails a cautionary tale for residents of Pyhäjoki, who seem to be well aware of the complexities behind and beyond their realm of action. Still, during the later stages of the siting procedure, the two Northern communities of Pyhäjoki and Simo were competing to become the new host for Finnish nuclear power production. Rather than being met with fierce opposition, the developers and their consultants witnessed demonstrations organized to convey support for the high-risk technology project. Accordingly, municipal leaders of Simo showed their disappointment while those of Pyhäjoki appeared satisfied as Fennovoima's decision was announced to them in front of the media. In colloquial terms, one could refer to this as a YIMBY, "yes-in-my-backyard" attitude (Lake, 1993), certainly a developer's dream situation (however, see McClymont and O'Hare, 2008, for a deconstruction of the pejorative terminology). The support for nuclear development was emphasized in ways that surprised the consultant firm facilitating hearings (interview with Pöyry Oy). Such support and the attitude of a local informant toward Fennovoima's nuclear power project (referred to at the beginning of the chapter) show how these two municipalities have collectively turned their relative powerlessness (see Blowers, 2010, for the British case) into a strong position for negotiation. The experiences from Eurajoki, Loviisa and other industrial projects and developments, not least in the mining sector, seem to have inspired these communities to take the chance and agree on high-risk technology in their vicinity for a number of tangible benefits: employment, revenues, industrial activity and political influence. The partnership between industrial developer and community decision-makers is actively sought and has become a part of the siting process that reaches far beyond what happens on the official stage (see Syrjämäki et al., 2015, for a critical analysis of public comments revealing anxiety and mistrust among residents about the soundness of the project). In a more formalized form, we can observe this in the context of negotiations for impact and benefit agreements of industrial development on indigenous territory in Canada, and the social license to operate in the mining sector (e.g. Lesser et al., 2017; Sosa and Keenan, 2001).

Nuclear facilities in the Arctic and sub-Arctic environments

In the initial stages of the siting procedure for Finland's sixth nuclear power plant unit, commenced officially in 2007, four other places had been included in the discussion: Vaala, Ruotsinpyhtää, Kristiinankaupunki (the only place where opposition was strong and loudly voiced) and Kemijärvi. When Fennovoima started to look for suitable locations, the town of Kemijärvi was struggling with the looming closure of its pulp mill. If the siting had gone ahead in Kemijärvi, it would have placed nuclear power production in the sparsely populated environment of the European Arctic (Arctic Circle boundary).

Even if Kemijärvi was soon dropped as a potential site for Finland's sixth nuclear power plant, the consideration of the remote Arctic town gives evidence of the desperate situation when another large employer was closing its facilities and thus created the need to attract large-scale business to the region. A nuclear power plant, as much as the many mines in operation and at the planning stage, however, conflicts with interests in avoiding an influx of nonlocal workers and goods in order to protect the environment and local people's livelihood based on this very environment. While already one of the most connected regions of the Arctic, Finnish Lapland continues to invest in infrastructure that can host increasing amounts of traffic by expanding road networks and airports and by discussing the connection of the railway to the Arctic Ocean. It remains a balancing act between extractive industry development and another major business, tourism, to encourage and at the same time limit expansion in order not to damage the area's precious but vulnerable resource, a rather pristine environment (Veijola and Strauss-Mazzullo, 2018).

A nuclear power plant in Kemijärvi would have been only 500 kilometers from the operating nuclear power plant in Murmansk, Russia. Russia has recently equipped the first floating nuclear power plant, the Akademik Lomonosov, for flexible use in Arctic waters: for instance, to help with the construction of oil and gas platforms. In a press release, Rosatom advertises the environmental benefits in an effort to reduce the Arctic region's dependency on coal (Rosatom, 2018). The low levels of greenhouse gas emissions from the nuclear fuel cycle is a frequently emphasized argument in Finnish debate. It is also a response to the decision by other European countries such as Germany to phase out nuclear power. Here, Finnish proponents of nuclear power argue that such a phaseout in central Europe is not possible without reverting to energy sources with high levels of greenhouse gas emissions, mainly coal. A sixth nuclear power plant unit in Finland, the necessity of which has been widely contested, is considered crucial by the proponents of stabilizing the Nordic and European electricity markets. On the other side of the Baltic Sea, Sweden has decided to eventually replace its operating nuclear reactors, and due to its high dependency on nuclear power, a phaseout is currently unlikely.

Furthermore, the current project by Fennovoima in Pyhäjoki is internationally exceptional and unprecedented because it means that the power plant will be built on a "greenfield site." It is a location that has not been industrially developed but, in fact, used to be an environmental heritage site in a place that is hundreds of kilometers away from existing nuclear power plants. The two competing municipalities of Pyhäjoki and Simo are also located in the vicinity of coastal cities and are not to be considered peripheral or remote, bearing in mind their location along major transport routes and infrastructure.

All in all, the developments in Finland appeal to managers worldwide who seek similar progress toward an expansion of nuclear power. Despite the events at Fukushima, a growing demand for renewable energies, scandals and cost overruns during the construction of nuclear power plants, growing public consciousness of the detrimental effects of uranium mining (abroad) and the difficulty of disposing of spent nuclear fuel safely – which, all together, entail a strong argument against nuclear expansion – Finland is the only democratic country to successfully site this contested high-risk technology for electricity production.

From nuclear legacies to future uranium mines

Radioactive material has long been present in the Barents region, including weapons testing, nuclear waste dumped in the Arctic Ocean north of Russia, sunken submarines equipped with nuclear generators and also the radioactivity that arrives through water and air currents, nowadays especially from Sellafield and La Hague (AMAP, 2016). Radioactive contamination of the Arctic is expected to rise also due to climatic changes, as thawing permafrost soils will release increasing amounts of radon. Inhabitants of the Arctic are more at risk of exposure to higher levels of radon because of Arctic food chains and people's diet. In addition, the faster melting of Arctic snow and ice accelerates the absorption of radon in water and soil, while its decay is still slower than in temperate climates.

The shipping of spent nuclear fuel from Finland to Russia stopped soon after 1994 when the decision was made to discontinue the practice. Finland established a timetable for how to deal with nuclear waste and promised to find a solution within an outlined schedule. The country kept to its schedule, unlike every other democratic country that has produced a similar timetable. The construction of the final repository has started at Onkalo, on the West coast of Finland, in close proximity to two existing reactor units (Olkiluoto 1 and 2, soon also 3). The spent fuel from the two nuclear power companies Fortum and Teollisuuden Voima will be stored away in the Finnish bedrock deep underground. The company responsible for dealing with nuclear waste, Posiva, has not allowed the more recently established Fennovoima to store its spent fuel in the same location. This remains a problem to be resolved.

Problems of knowing and societal solutions

As yet unresolved are concerns over the beginning of the fuel cycle. The last uranium mine closed in 1961 in East Finland, at Paukkajanvaara. The site has subsequently been restored and thus provides an "interesting field laboratory" (Colpaert, 2006) to study radioactive emissions from a closed mine site. Colpaert's terminology sketches the problem inherent to high-risk technologies as well as the production and disposal of its (used) resources. Where activities and their effects cannot be tested under safe conditions in a laboratory, we mostly rely on field experiments, drawing the whole society into the experiment (Krohn and Weyer, 1994). This act requires additional legitimacy beyond scientific motivation as it puts people and the environment at potentially greater risk of suffering from yet-unknown consequences due to complex relationships in the natural and human environment. The storage of (highly radioactive) spent nuclear fuel in the bedrock is one example where uncertainties over natural processes and human activities are beyond comprehension because of the project's extended time frame.

This further illustrates a significant problem inherent to science as described by seminal authors. Only identified knowledge gaps can be investigated through experimentation in the laboratory or through careful and restricted implementation in the field. A famous example is field experiments with genetically modified (GM) crops, which are carefully monitored at the European level, however with an awareness that unintended effects may be irreversible (Levidow, 2001). Yet the biggest risk for the natural environment and its human inhabitants lies in unidentified knowledge gaps (*unspezifisches Nichtwissen*) (Japp, 1999; Luhmann, 1990), those areas of which scientists may be completely unaware.

The actual discovery of radiation by physicists illustrates how lack of awareness of its damaging effect on human cells only became evident when scientists fell sick and died from exposing themselves. Marie Curie established a theory of radiation following experiments with radium and polonium and later fell sick and died from prolonged exposure. Once increasing numbers of people fell sick and a potential connection to the unprotected handling or voluntary exposure to radioactive material was established, the unspecific knowledge gap turned into a specific one. The list of similar examples runs through the history of science.

The political "solution" to ensure safety is to burden the implementation of field experiments with additional requests for evidence on potential hazardous effects as well as precautionary measures. However, these can only target identified knowledge gaps (known uncertainties), whereas unidentified knowledge gaps may only become known areas of uncertainty when the damage has reached a dangerous scale on an irreversible path.

Where knowledge is identified to be lacking but is no longer pursued once certain standards have been established, we can also argue for the prevalence of "negative knowledge" as defined by Knorr Cetina (1999), describing the understanding that additional research in that specific direction is not necessary as the additional knowledge gain is not perceived to lead toward a major breakthrough. Such cases of "ignorance" have become a frequent criticism in the siting of high-risk technologies and also in medical sciences and the allocation of research funds. Incidences of cancer around nuclear facilities are often claimed to be lacking attention by scientific and government institutions. The notion of unspecified uncertainty, however, entails a universe of potential problems of which society is yet completely unaware.

The social solution to the problem of not knowing the risks involved is partly one of compensation (Kojo, 2008), which builds on the possibility that nothing will happen as the history of nuclear power management in Finland suggests. Thus, the Finnish field experiment with nuclear power may be considered fairly established routine for a range of societal actors and benefits from the absence of major incidents over the course of almost fifty years. Finland does not suffer from earthquakes and is not the target of terrorist attacks and rising sea levels are canceled out by a continuous land uplift.

In our observations of attitudes and events in Pyhäjoki and Eurajoki, we witness how residents have normalized the exceptional situation of (soon) living close to nuclear power plants in Finland. Almost fifty years of operation in Loviisa and Eurajoki have passed without major incidents. Although this is a short moment in terms of radioactive half-life, it almost extends over a human life span.

Uranium as a domestic product

Looming scarcity and rising prices in the oil and gas sector as well as concerns over greenhouse gas emissions have renewed interest in not only nuclear power, but also the domestic production of uranium and other radioactive substances, radium foremost. In addition, an ethical consciousness is on the rise that claims greater responsibility for the consumption of goods and resources from abroad, especially if those come from poor and dependent regions. Uranium and radium are not excluded from this discourse, especially in communities that have long experiences with the extraction of minerals and thus have previously granted a social license to operate (Litmanen et al., 2016). Since the 1990s, an increasing number of mining companies have searched for the resources all across Finland and mapped deposits in the bedrock. Amendments of mining regulations and a favorable public attitude (not overly positive, but less negative than in other democratic countries) prevailed – up to a certain point. The course of events at the mine in Talvivaara at Sotkamo in Eastern Finland has eroded public support of any such activity since 2010 (Heikkinen et al., 2016). The lack of communication with the public and the surfacing of information about environmental damage leading to a criminal investigation have had an impact on all mining activities in Finland, claim Sairinen et al. (2017). While the handling of nuclear power production as well as the siting of the final storage for spent nuclear fuel have been settled within a (comparatively) strong atmosphere of trust in responsible authorities and democratic institutions such as the decision in principle, the mining of radioactive material is now connected to an image of ruthless and corrupt private management, despite a generally favorable public attitude toward domestic production of goods and resources.

Because of continuous opposition to uranium/radium mines, the material is often harvested as a byproduct, as in the case of Talvivaara in Finland. In any case, it usually takes decades from the exploration license to the actual operation of a mine. Across the European Union, no active uranium mine can be found (the Czech Republic is closing its facilities), with some countries such as Sweden holding a ban on uranium mining. Greenland lifted its ban in 2013 and is now investigating the feasibility of a mine in Kvanefjeld in South Greenland. Thisted (2018) has interpreted the attitude of the Greenlandic political elite in power. In reference to Ahmed (2010), she proposed that uranium be conceived of as a "happy object," a tool for Greenlandic society to become economically independent from its former ruler Denmark. Closer proximity to an object (and, I would say, independence from Denmark) create expectations of a future state of being. Happiness lies ahead if one does the right thing.

Discussion

Taking this notion in the Finnish context means understanding nuclear power production and the final disposal of nuclear waste but also past and potential uranium mines as part and parcel of a content nation. In popular polls, Finland ranges at the top of happy nations (Helliwell et al., 2018). According to Ahmed (2010), happiness is an expression of knowing that things are being done the right way and that they will lead to a desired future. That future is at risk from climatic changes, social inequalities, out-migration and the increasing dependence of small northern communities. In this context, nuclear power production has appeared as a lesser evil to some communities in the Finnish North in its Arctic and sub-Arctic environments. Beyond the local community, it provides a relatively easy and already-established path toward maintaining good living standards. In addition, an identity is taking shape that claims benefits without burdening others more than necessary. Thus, the ethical consciousness that makes nuclear power expansion possible in this unprecedented form (a new nuclear power plant on a greenfield site, the construction of a final repository for spent nuclear fuel, and the license to harvest uranium at Sotkamo) is one that prefers the domestic production of goods and the domestic disposal of waste produced during consumption. It ideally includes the whole fuel cycle of the high-risk technology, from the exploitation of the resource to its "final" storage, despite the fact that the enrichment and actual fuel production would require the raw material to be sent abroad as there are no enrichment plants located in Finland. In a sense, nuclear power production can be conceived of as a tool toward happiness, although not a conscious one as in the Greenlandic case; it is embedded within a content attitude. And the Finnish case has similar motives toward independence: we can observe this in the careful maneuvering between safety standards and influence from the former ruler and superpower Russia (Haukkala, 2018) and also in the argument to become independent from electricity imports through the Nordic electricity market.

While we can observe the overall tendency to support nuclear power production in a land of few resources (mainly water, peat, timber and, more recently, wind – all of them more or less contested), a closer look reveals different attitudes, divided communities (Kari et al., 2010) and low degrees of civic involvement. In order to understand the lack of organized protest against high-risk technology megaprojects in Finland, I have considered assumptions about the role of the "strong administrative state" (Säynässalo, 2009, see earlier) that favors decision-making by a technocratic elite with close industry relations, at the same time leaving little room for civil society (Litmanen et al., 2017, see earlier). Ruostetsaari (2018) further describes the low degree of civil efficacy (third lowest in comparison to 22 European countries) in Finland as a form of political consumerism, citing data from a 2008 survey. This goes along with high degrees of trust in democratic institutions, although not in political parties (Suomen vaalitutkimus, 2016). In the nuclear case, this includes the governmental "decision in principle" on the question of whether the proposed project is in line with the "overall good of society" (Nuclear Energy Act 990/1987; Strauss, 2011).

Concluding remarks: uncertainty and contentment – living with nuclear

Will Pyhäjoki in Northern Finland become another proud "electric" community like Eurajoki in the Southwest of Finland? While progress in the preparation of the next nuclear facility site is slow, things are going forward. The community of Pyhäjoki is keeping a low profile, generally supporting the megaproject. Protest is scattered and met with local opposition, as the construction of Finland's sixth nuclear power plant means economic activity, employment and revenues for the municipality. In response to the question of whether he feared a nuclear power plant in his neighborhood, the Pyhäjoki resident quoted earlier expressed a locally common attitude: "if there was a nuclear accident in our country, we would be affected anyway," claiming that the price would be paid by those living both near and far.

For the people of Pyhäjoki, the resource has multiple facets, many of them positive, starting with the fact that the community has become a place on the map, beyond national borders. On the ground, a public-private partnership is growing, as public buildings and infrastructure receive additional funding and events are sponsored. In any case, "living with nuclear" is far from becoming actuality any time soon as the construction can easily take decades. In contrast to the uncertain risk, the resources the community can benefit from now are visible, even tangible. In similar fashion, the "most electric" community in Eurajoki has decided to support the construction of a final repository, thus being able to benefit from the final storage through taxes. For the time being, power production and final storage have been resolved socially. The mining and enrichment of uranium, however, remains a very different story, not least because of the unfortunate events and attempted cover-ups at Talvivaara mine in Eastern Finland.

In this chapter, I have argued that the production of nuclear power functions as a tool to reach or maintain a state of contentment in Finnish society, despite the technology's inherent and unresolvable uncertainties and dangerous scales. Independence or self-sufficiency (Haukkala, 2018) is a major goal in Finnish energy policy, combined with the effort to reduce emissions and to deal with waste in the domestic context. Recently, the extraction of uranium has also been considered and began to be implemented. Such a development is not possible without the willingness of communities to carry the burden of living in close proximity to high-risk technologies. Observations from different Finnish municipalities help examine high degrees of acceptance and trust among residents and enable us to understand the degree to which communities are divided and contest the role of decision-makers and safety watchdogs.

References

Aalto, P., Nyyssönen, H., Kojo, M. and Pal, P. (2017) "Russian nuclear energy diplomacy in Finland and Hungary," *Eurasian Geography and Economics*, 58(4), pp. 386–417.

Ahmed, S. (2010) *The promise of happiness*. Durham: Duke University Press.

AMAP (2016) *Radioactivity in the Arctic*. Oslo: Arctic Monitoring and Assessment Programme.

Blowers, A. (2010) "Why dump on us? Power, pragmatism and the periphery in the siting of new nuclear reactors in the UK," *Journal of Integrative Environmental Sciences*, 7(3), pp. 157–173.

Choi, Y. (2018) "Trust in nuclear companies and social acceptance of a nuclear waste repository in Finland," *Journal of Environmental Information Science*, 2018, pp. 44–55.

Colpaert, A. (2006) "The forgotten uranium mine of Paukkajanvaara, North Karelia, Finland," *Nordia Geographical Publications*, 35(2), pp. 31–38.

Haukkala, T. (2018) "A struggle for change: the formation of a green-transition advocacy coalition in Finland," *Environmental Innovation and Societal Transitions*, 27, pp. 146–156.

Heikkinen, H.I., Lépy, É., Sarkki, S. and Komu, T. (2016) "Challenges in acquiring a social licence to mine in the globalising Arctic," *Polar Record*, 52(4), pp. 399–411.

Helliwell, J.F., Layard, R. and Sachs, J.D. (eds.) (2018) *World happiness report*. New York: Sustainable Development Solutions Network.

Japp, K.P. (1999) "Die Unterscheidung von Nichtwissen [The distinction of uncertainty]," *TA-Datenbank-Nachrichten*, 8, pp. 25–32.

Joona, T. (2018) Personal communication with Tanja Joona, a researcher in the ongoing project "Wollie—Live, work or leave? Youth wellbeing and the viability of (post) extractive Arctic industrial cities in Finland and Russia" at the University of Lapland.

Kari, M., Kojo, M. and Litmanen, T. (2010) *Community divided: adaptation and aversion towards the spent nuclear fuel repository in Eurajoki and its neighbouring municipalities*. Final Report of SEURA Research Project. University of Jyväskylä & University of Tampere.

Knorr Cetina, K. (1999) *Epistemic cultures: how the sciences make knowledge*. London: Harvard University Press.

Kojo, M. (2008) "Compensation as means for local acceptance. The case of the final disposal of spent nuclear fuel in Eurajoki, Finland," Paper presented at the conference Waste Management, February 24–28, Phoenix, AZ.

Krohn, W. and Weyer, J. (1994) "Society as a laboratory: the social risks of experimental research," *Science and Public Policy*, 21(3), pp. 173–183.

Lake, R.W. (1993) "Planners' alchemy transforming NIMBY to YIMBY: rethinking NIMBY," *Journal of the American Planning Association*, 59(1), pp. 87–93.

Lesser, P., Suopajärvi, L. and Koivurova, T. (2017) "Challenges that mining companies face in gaining and maintaining a social license to operate in Finnish Lapland," *Mineral Economics*, 30(1), pp. 41–51.

Levidow, L. (2001) "Precautionary uncertainty: regulating GM crops in Europe," *Social Studies of Science*, 31(6), pp. 842–874.

Litmanen, T., Jartti, T. and Rantala, E. (2016) "Refining the preconditions of a social licence to operate (SLO): reflections on citizens' attitudes towards mining in two Finnish regions," *The Extractive Industries and Society*, 3(3), pp. 782–792.

Litmanen, T., Kari, M., Kojo, M. and Solomon, B.D. (2017) "Is there a Nordic model of final disposal of spent nuclear fuel? Governance insights from Finland and Sweden," *Energy Research & Social Science*, 25(Suppl. C), 19–30.

Luhmann, N. (1990) *Die Wissenschaft der Gesellschaft* [The Science of Society]. Frankfurt am Main: Suhrkamp.

McClymont, K. and O'Hare, P. (2008) "'We're not NIMBYs!' Contrasting local protest groups with idealised conceptions of sustainable communities," *Local Environment*, 13(4), pp. 321–335.

Müller, W.C., Thurner, P.W. and Schulze, C. (2017) "Conclusion: explaining nuclear policy reversals" in Müller, W.C. and Thurner, P.W. (eds.) *Politics of nuclear energy in Western Europe*. Oxford: Oxford University Press, pp. 286–322.

Official Statistics of Finland (OSF) (2017) *Energy supply and consumption* [e-publication]. Helsinki: Statistics Finland. Available at: www.stat.fi/til/ehk/2017/ehk_2017_2018-12-11_tie_001_en.html (Accessed: April 9, 2019).

Parkhill, K.A., Henwood, K.L., Pidgeon, N.F. and Simmons, P. (2011) "Laughing it off? Humour, affect and emotion work in communities living with nuclear risk," *The British Journal of Sociology*, 62(2), pp. 324–346.

Rosatom (2018) *Floating nuclear power unit Lomonosov has arrived in Murmansk to be loaded with fuel* [Online]. Available at: www.rosatom.ru/en/press-centre/news/floating-nuclear-power-unit-lomonosov-has-arrived-in-murmansk-to-be-loaded-with-fuel-/ (Accessed: April 9, 2019).

Ruostetsaari, I. (2018) "Political consumerism as a means in influencing energy policy and solving environmental problems. The case of Finland in 2007–2016," *International Journal of Economy, Energy and Environment*, 3(3), pp. 21–31.

Sairinen, R., Tiainen, H. and Mononen, T. (2017) "Talvivaara mine and water pollution: an analysis of mining conflict in Finland," *The Extractive Industries and Society*, 4(3), pp. 640–651.

Säynässalo, E. (2009) "Nuclear energy policy processes in Finland in a comparative perspective: complex mechanisms of a strong administrative state" in Kojo, M. and Litmanen, T. (eds.) *The renewal of nuclear power in Finland*. London: Palgrave Macmillan, pp. 126–160.

Sosa, I. and Keenan, K. (2001) *Impact benefit agreements between aboriginal communities and mining companies: their use in Canada* [Online]. Toronto: Canadian Environmental Law Association, Environmental Mining Council of British Columbia, CooperAcción. Available at: www.cela.ca/sites/cela.ca/files/uploads/IBAeng.pdf (Accessed: April 9, 2019).

Strauss, H. (2011) *For the good of society: public participation in the siting of nuclear and hydro power projects in Finland*. PhD thesis, Acta Universitatis Ouluensis, Oulu.

Suomen vaalitutkimus (2016) *Luottamus yhteiskunnallisiin instituutioihin* [Trust in public institutions, Online]. Available at: www.vaalitutkimus.fi/fi/poliittiset_asenteet/luottamus_yhteiskunnallisiin.html (Accessed: April 9, 2019).

Syrjämäki, E., Kojo, M. and Litmanen, T. (2015) *Muuttunut hanke. Fennovoiman ydinvoimalahankkeen YVA- yleisötilaisuudet Pyhäjoella vuosina 2013–2014*. Jyväskylä: YFI Publications.

Thisted, K. (2018) "Greenlandic exceptionalisms," Paper presented at the Northern Political Economy Symposium: Arctic Continuities, August 29, Arctic Centre, University of Lapland, Rovaniemi.

Veijola, S. and Strauss-Mazzullo, H. (2018) Tourism at the crossroads of contesting paradigms of Arctic development" in Finger, M. and Heininen, L. (eds.) *The GlobalArctic handbook*. Cham: Springer, pp. 63–81.

YLE News (2018) *Russia threatens counter-measures if Finland and Sweden join Nato* [Online]. Available at: https://yle.fi/uutiset/osasto/news/russia_threatens_counter-measures_if_finland_and_sweden_join_nato/10321784 (Accessed: April 9, 2019).

4 Untied resource as a threa/-t/-d for social fabric(ation)

Joonas Vola

Introduction

Life in Finnish Lapland in the Arctic is entangled around water bodies that cover people's lives' requirements. As natural transport corridors, rivers have brought the marketplace into their crossings while also serving as sources of local natural subsistence. In the wake of industrialization and modernity, they have become resources for hydropower. Bodies of water are the source of life and prosperity where the actual linguistic heritage of the resource is embedded in the river. The economy may be a source for societal life, but it can develop to such an extent that it both challenges ecological sustainability and becomes unbearable for social sustainability. Social responsibility may require local sacrifices to benefit society at large, but this approach may dismiss place making and community development. An ecologically flourishing river may also emerge as a risk factor for the surrounding community if the infrastructure makes inroads into the natural dynamics of the river. Sustainable development along the river is a line drawn in the water.

This study dissects the genealogy of language entitling the river as a resource. I claim that the materiality of lives, both of the river and of the inhabitants alongside it, are spun in the language in rationed, calculable, predictable and binding lines that are inseparably entangled. The language that is based on the facts and fixed meanings, therefore, both brings and requires predictability to and from the materiality of the world, producing risk prevention of natural hazards and utilizing natural resources. These measures enable social order in the area and, therefore, also subject the inhabitants to the resource language. The study contributes to the common resource-driven Arctic-oriented studies by problematizing the used language itself. It offers a new opening to a long and continuing regional paradigm on resource use, the case of the Kemijoki (Kemi River) water reservoir.

The chapter builds on the principles in Artur Przybyslawski's study on the language of becoming (Przybyslawski, 2003) and is inspired by Michel Foucault's genealogical approach to discourse analysis that recognizes the historical materiality and power relations invested in the current practices (Foucault, 1977). The analysis utilizes etymology to break down the key terms to open up their relatedness and materiality. The linguistic heritage and metaphors are disentangled from the current discourse on the debate around the management of the river, to recognize political rationalities leading to certain aims, objectives, controversies and disputes.

The analysis starts by describing a way of speech where a spoken entity, a natural resource, is put into terms of calculable units, using the Kemi River management as a case study. It is followed by recognizing the linguistic connection between the meaning and materiality of the river and the domination of the language of potentiality in contrast to unstructured contingency, which is conceived as an impossibility for the social order. The discourse evolves to the langue of risk and progress, where measures for flood risk prevention and increased hydropower come together in reemerged water reservoir plans. This is where the requirement of predictability as the fundamental basin for social order is studied in terms of classical approaches to the relation between social order and economy. Due to the legal, administrative and communal borderlines, the case changes from an issue of environmental engineering to a question of regional development and cohesion of social order, where social sustainability and economic growth are interlinked. The findings are generalized in a philosophical inquiry by examining the language in relation to deep, cultural metaphors: the bow (pienellä alkukirjaimella, näin se on jo valmiiksi jäljempänä) of Heraclitus and the thread of life. These represent linear thinking and identity built on controversies, which lead to an impression of predestined future and exclusive possibilities. The correlation between the generalized findings and specific literal statements are examined by analyzing the recent public debate on the Kemi River management after the latest disputes and decisions made in spring 2018 by using the key metaphors as an interpretative and constructing frame for the conclusions.

River as a re/source

In terms of political economy, resources count as "a country's wealth." As a term, "resource" often becomes meaningful when coupled with a pretext, or a word such as "natural" or "human." What the term actually does not indicate is the actuality of possessing or containing something – only the possibility of such outcomes. In the 1610s, the term indicated the "means of supplying a want or deficiency," from the French resourse, tästä on pudonnut pois 'r' kirjain, kun tarkistin lähteen (vuoden 1610 kirjoitusasu), that which is not yet supplied but where there is a want or need as well as a means to do it. The etymological root in Latin, *resurgere*, translates as "rise again," where the prefix "re-" commonly expresses something that is done again or repeated as in the words "reoccur" and "represent" (Online Etymology Dictionary, 2018a). The following part, "source" (n.), has its history in the Old French *sourse* as "a rising, beginning, fountainhead of a river or stream" and derives from Latin *surgere*, "to rise, arise, get up, mount up, ascend" (Online Etymology Dictionary, 2018b). Therefore, to physically locate and point out a source would to be discover the fountainhead of a river. The term "resource" also has another relativity: *surrigere*, from an assimilated form of *sub* ("up from below"), combined with *regere* ("to keep straight, guide"), meaning "to direct in a straight line," thus, "to lead, rule" (Online Etymology Dictionary, 2018a). The resource and directing are therefore connected in language. One is the starting point of the stream, and the other is the directing line.

The studied river, as far as facts and figures go, can be describes as follows (see Map 4.1). The Kemi River, at 550 kilometers, is the longest river in Finland, crossing the cities of Kemijärvi and Rovaniemi, until it reaches the sea between Kemi and Keminmaa with an average flow of 556 cubic meters per second. Its catchment basin of 51,000 square kilometers covers most of Northern Finland and extends over the border to Russia. The tributaries of Ounasjoki (Ounas River) and Raudanjoki connect to the Kemi River in the area of Rovaniemi, while the Kitinen and Vuotosjoki find their connections upstream in Pelkosenniemi and Tenniöjoki in Savukoski. Such a description gives the river a course: beginning, end, direction, length, volume and order in relation to other bodies. When the running of

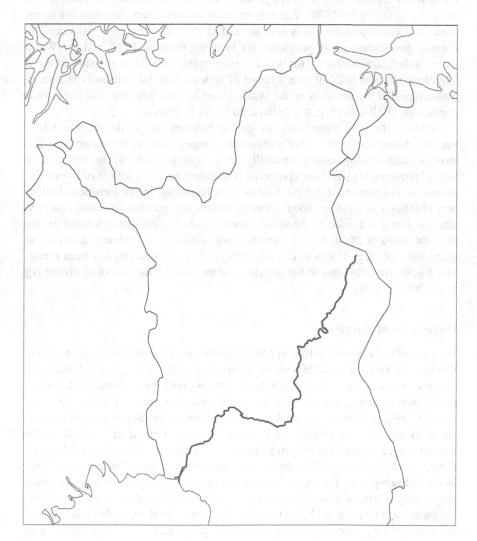

Map 4.1 Kemi River, Finnish Kemijoki

the river is put into figures, as measurements, averages and estimations, it is also a step toward a discourse where the waters can be managed as numbers and calculable units, apart from the actual flow. It translates the pure contingency of the river into a language of potentiality, a resource talk.

Speech materializes as a resource as follows. The first hydropower plan in the Kemi River's catchment was built at Isohaara in 1946; there are now 21 hydropower plants in its catchment areas. Kemijoki Ltd., established in 1954, owns 16 of them. The average energy production in the catchment area is 4,124 gigawatt hours per year (Kemijoki Ltd., 2018). As the Ounas River was protected in 1983, it has remained without hydropower plants (ELY Centre Lapland, 2003). The Kemi River branch reservoir plan's predecessor, the artificial lake of Vuotos, was intensely driven in the 1970s. The plan to build a water reservoir where the Kemi River and Vuotosjoki meet was first aborted in 1982 due to civic resistance. The Finnish government again promoted the building plans in 1992, and the Water Court, a dedicated authority for water management, granted permission to build a hydropower plan in 2000 after a round of appeals. The Administrative Court of Vaasa repealed the decision of the Water Court in 2001 and rejected the permit application (KHO, 2017, p. 87; Laitala, 2013; YLE, 2011).

Paradoxically, this prosperous life-giving prospect has as its obverse a life-limiting power due to the river's seasonal changes, such as the threat to infrastructure and housing caused especially by the spring floods. Therefore, another way of pronouncing the river's potential is to calculate it as a source of risks to be estimated and prevented and the cost of such measures. In the European Union, tens of billions of euros are spent annually to mitigate, recover from and compensate for damages caused by flooding, especially during the difficult flood events in 2000 through 2009, where damage control mainly concerned urban areas (Keskitalo, 2013, p. 1; van Ree et al., 2011, p. 874). Rovaniemi has been recognized as such an area due to the influence of the Kemi River flooding (Tennberg et al., 2018, p. 209).

Dis/course of the river

If I were to write down the title as *Untiedresourceasathreatdforsocialfabrication*, it would be an unreasonable spur of letters that might or might not hold some meaning, but as such, it has no use. If the title is divided into words and if some conditional markings are added, it becomes *Untied resource as a threa/-t/-d for social fabric(ation)* and presents different possibilities for the outcome of the sentence. Where the first example is pure contingency without determination, the second has the potential to bear two outcomes, one or the other, structurally interlinked. Suvi Alt (2016, p. 119) defines the terms "contingency" and "potentiality" in the following way: contingency is what *may but need not be or happen*, while potentiality is what *will or ought to come into being*. Therefore, "contingent" has the capacity but not the necessity to be in one form or another, while "potentiality" is not potentiality without a certain pressing urgency; its presence is defined by its future. Contingency would allow it to be forgotten, potentiality only to be delayed.

According to Mark Haugaard's analysis, a state of pure contingency would undo the possibility of social power. The added capacity for agents' actions gained from society derives from the existence of social order: in other words, from the predictability of social life (Haugaard, 2003, p. 93). Power, as "being able to" (Online Etymology Dictionary, 2018c), is relational because it takes place in relation to something. Therefore, power returns to the social order. Haugaard, building on the work of Arendt, Barnes, Giddens, Luhman and Parsons, claims that the creation of social order is the precondition for power as an outcome. Instead of coercion, a blend of actors reproduce social order, a society, constraining and facilitating the capacity for action (Haugaard, 2003, p. 88). Agents derive the capacity from nature and society. Capacity to act comes from the physical body and from the ability of humans to harness the laws of nature to their advantage. The latter form of natural power presupposes knowledge of nature's regularities that manifest in cause and effect. Even a mistaken knowledge may be found sufficient if it allows any predictability of the phenomenon (Haugaard, 2003, p. 89). Structures create power through predictability, therefore precluding certain forms of interaction, organizing into politics and organizing out. Social order is a question of structuration, confirming structuration and disconfirming structuration (Haugaard, 2003, p. 96). Life without social order would emerge as socially unsustainable.

The way in which social, order and power co-constitute the world also goes with a discourse. The word *course* (n.), from the 13th century, means "onward movement, path or distance prescribed for a race" and comes from the Old French term *cors*: "running; flow of a river," originating from Latin *cursus*: "a running; a journey; direction, flow of a stream." In the 13th century, it also meant "order, sequence" (Online Etymology Dictionary, 2018d). To summarize, the word means movement in a predictable or predicted location: a runner on a track, a river in its basin or a boat in the sea. By adding the prefix "dis-" (apart) to "course", discourse (n.), from the late 14th century, stands for a "process of reasoning," referring to "a running over a subject in speech, communication of thought in words" from the 1550s onward (Online Etymology Dictionary, 2018e). In short, I claim it to be a practice where a being, a thing or a phenomenon is treated with means and matters of language. Therefore, it has given a subject to speech while being subjected by speech. Discourse is not independent from the substance that it is "running over" but can continue to move "apart" from the substance's course. Therefore, the language of the river may flood the embankments, which still hold the water.

"Reservoir" has been known since 1705 as an "artificial basin to collect and store a large body of water," derived from the French word *reservoir*, a "storehouse," from Latin *reservare*, to "keep back, save up; retain, preserve," from the Proto-Indo-European root "ser-," "to protect" (Online Etymology Dictionary, 2018f). What is "protected" here? Besides protecting the community's access to the resource, it serves to protect the community from being overrun by the waters. Therefore, water management and adaptation are taken into consideration in numerous legal instruments, such as the Water Framework Directive (2000/60/ EC), the Floods Directive (2007/60/EC) and the Water Scarcity and Drought Communication (Keskitalo, 2013, pp. 18–19). The EU Water Directive (2000), a predecessor of the EU Flood Directive (2007), subjected all types of flooding to a scheduled planning for prediction and prevention: preliminary flood risk assessment and identifying major flood-prone areas by 2011, related flood hazard and risk mapping by 2013 and finalizing flood-risk management plans by 2015 (Tennberg et al., 2018, p. 209). The mapping project was funded by the EU (Keskitalo et al., 2013, p. 88), and within this process, Rovaniemi was recognized as a flood risk area due to the Kemi River, an Arctic river seasonally covered by ice and snow, which is estimated to be influenced by climate change in relation to the volume of spring floods. In 2002, after the Supreme Administrative Court repealed the plan for the Vuotos water reservoir, the Regional Council for Lapland started to plan for a smaller water reservoir (Rytkönen, 2018, p. A4) that would also function to prevent flooding (Lapin Kansa, 2018, pp. A4–A5). An area was thus carved out for a water reservoir in the 2016 provincial plan of Eastern Lapland (Lapin Kansa, 2018, p. A2). This plan was abandoned by the Finnish government on February 1, 2018, and the measures for flood protection have remained unsettled (Lapin Kansa, 2018, pp. A1–A2).

Eco/nomy of the river

In Finland, the local government is granted self-determination in planning (Keskitalo, 2013, p. 3), which means that the flood preparations are also influenced by local development aims along with national requirements and EU directives. In another flood-risk area in Finland, the Tornio River, a flood risk exists, but it is relatively in terms of the level of preparedness and the scale of damages. The adaption of the floods directive can be seen as an external policy event that has come to exert an impact on the local situation while the development pressures in the area potentially encourage risky building close to the water (Keskitalo et al., 2013, pp. 88–89). To put it very roughly, the local emergency management is proceeding with the adaptation strategies due to the enforcing regulations of climate change adaptation rather than because of a pressing change in the actual hydrological and climatological conditions requiring adaptive measures: they are adapting to the flood directive rather than to the flood. At the same time, some risks are built into the city planning and therefore require adaptation measures. Where the ecological and social development do not sustain the argument for the required measures alone, the focus needs to be moved to the economic development instead.

Overspills during the spring floods have been calculated to cause losses of as much as 40 million euros annually for the hydropower companies' energy production because the water surplus cannot be used as a reserve for the dry winter season (Laukkanen, 2013). This line of argumentation calculates "loss" as a loss of potential profit, not as actual financial loss. In other words, the lost 40 million euros did not exist in the first place but could have existed. Such evaluation of water strongly indicates Karl Marx's understanding of the capitalist social system where objects become metamorphosed into commodities of exchange value, in contrast with the pre-capitalist world, where the object's meaning was defined in terms of use (Haugaard, 2003, p. 93). Therefore, for the hydropower company, unused resources equals loss.

In 2016, the spring flood and the rainy summer season improved the average production of Kemijoki Ltd. by 20 percent. While the flow of water was close to the average of a flood, Kemijoki Ltd. promised to lower the level of Lake Kemi-järvi during the autumn (Talvitie, 2016). As the water reserve relates to the accumulation of capital, according to the only norm of capitalism, it is to be invested due to the primary objective of self-expansion (Ruuska, 2017, p. 56). In the case of the Kemi River, this means the expansion of the total water reserve of the river to its limits, since the water has become potential capital. If the higher accumulation of water causes a higher risk for flood, it can be considered a risk investment.

Increased electricity production by the Kemi River reservoir would mean 0.5 percent of the total production within Finland and 6.5 percent for Kemijoki Ltd. The increased income to the region from municipal, real estate and corporate taxes is estimated to be between 70 and 84 million euros divided to the following 30 years and include 2700–2900 man-years amount of work. Of the annual increase of 300 GWh, 130 GWh would come from the new hydropower plant in the river branch and 170 GWh from the old plants along the main river; 60 percent of the increase would thus be made elsewhere than in the municipalities where the reservoir would be built (Kivelä, 2016). The area in which the energy industry invests does not receive equal advantage from the profits, even though it carries the socioecological risk. This indicates weak social sustainability and resembles what Albert O. Hirschman calls a "self-destruction thesis," a tendency of capitalism to destroy the social and moral foundations on which it rests (Kangas, 2009, pp. 301–302). The societal impact is evident in the current case, where the regional interests and the local interests do not meet and therefore cause continuous disagreement between the neighboring municipalities along the river. In contrast to Adam Smith's claim that the lowered political involvement and loss of martial spirit are due to the increase of commercialism (Kangas, 2009, p. 296), political involvement is high in the presented case, where both the city of Rovaniemi and the Regional Council of Lapland have promoted the reservoir project.

My findings show how the contingency of the river has turned into a potentiality, either as a potential resource for regional development or as a potential risk for the societal infrastructure and securing people's lives. Whether one or the other will come to actuality, the conclusion is development or a disaster. Furthermore, the risk of loss is not purely a collective one but has been more closely bound to individual risk carrying. The state compensation for flood damages, which covered a maximum of 80 percent of the damage caused by a 20-year flood, was replaced and "privatized" in 2014 with a system where property owners have insurance against 50-year floods as part of the property insurance offered by private insurance companies. Minimum flood insurance is mainly included in the compulsory real estate insurance. For better compensation, the insurance companies require that the property be located in a protected non-flood-prone zone (Tennberg et al., 2018). Here the current understanding of economy and the etymology of economy as "household management" (Online Etymology Dictionary, 2018g) meet.

βεcómíng of the river

The line between the risk and the profit, for socially and economically sustainable development in the region, is narrowing. For a political solution, the language of potentiality seems biblically to follow Matthew 7:13–14: to enter life proper is to follow the narrow line that leads to life, where the broad road leads to destruction. Most Western languages are linear due to basic sentence construction, encouraging speakers to focus on linear causal relationships rather than circular or mutually causative ones (Anderson and Johnson, 1997, p. 17). In the Western history of imagination, a line and a bow are common metaphors for life: a lifeline and a lifespan. They illustrate life as a temporal phenomenon that has a rational attempt to develop after the fixed starting point, reaching its highest peak and coming to an end, events following each other in a line. Metaphors are pervasive in everyday life, not just in language but fundamentally in terms of what we both think and act (Lakoff and Johnsen, 2003, p. 4). These ordering and predicting guiding lines, apparent in different diagrams and mapping, weave together hydrological models, risk reduction, electricity production and economic development models, indicating volume and time or, in different terms, the quality of living and the length of life.

The paradox of living a life, a narrow path, is interwoven with the heritage of language in European history. Ancient Greek had two words, βίος and βιός, separated only by an emphasis, in their literal form signified with the diacritic acute accent (´). In meaning, they appear to be very distant from each other. This paradoxical coexisting contradiction was highlighted by the ancient philosopher Heraclitus: "The name of the bow is life; its work is death" (Przybyslawski, 2003, p. 156). According to the etymological study, βίος originates from a Proto-Indo-European root, meaning "alive" (PIE reconstruction: $*g^wih_3wós$) or "to live" (Wiktionary, 2018a), whereas βιός comes from Proto-Indo-European ($*g^wiH$-) and cognates with Sanskrit '*jiyā́*', meaning "bow-string" and Old Church Slavonic 'žila', "vein, tendon" (Wiktionary, 2018b). The linguistic analysis appears to arrive at the forms of "alive" and "vein, tendon," one an abstract term and the other a concrete object, but it does not show their connection without the practical implication of the words and the historical order of things in the world of antiquity.

Artur Przybyslawski studies the paradigm of the "bow of Heraclitus" as a language of becoming in the following way. Przybyslawski says that naming affirms the existence and that one way to confirm the accuracy of naming is to realize the function of the named entity, and see if there is any contradiction. Then again, Przybyslawski presents how Heraclitus problematizes this approach. Instead of harmony, Heraclitus suggests that the functionality might be based on the very nature of contradiction within the entity. Names "do not comprehend how a thing agrees with itself." Heraclitus presents two concrete instruments, a bow and a lyre, which not only agree with themselves but also are at variance with themselves. As a result, they lack identity and remain lifeless (Przybyslawski, 2003, p. 155). It is the use or practice which makes things/beings what they are: the bow is a bow only when stretched by the archer when the bow is at variance, not agreeing with itself (Przybyslawski, 2003, p. 156). Gilles Deleuze claims that for Heraclitus names affirm only one side of reality, forced by naming to stop universal flux (Przybyslawski, 2003, p. 156). A bow, in principle, contains an act. It is not straight/straightened or bent/bend, but bow/bowing. Therefore, it should be treated as a verb rather than as a noun. It indicates momentum from one position to another.

The existence of a bow has another paradigmatic feature as well. The use of the bow affirms its identity but simultaneously risks its being because what is bent might as well be broken. In Przybyslawski's words the "usage changes the bow into something different and that usage contains the possibility of its damage, its destruction" (Przybyslawski, 2003, p. 156). This is the point where we need to return to the words βίος and βιός. Where are these two words interwoven? A tendon in the bow originates from an organism. It has bound the muscles to the skeleton, power and structure together. To gain a tendon, it needs to be extracted by taking a life. The aim of a hunting bow is in a living, and its goal is killing. The killing, on the other hand, is a necessity for nurturing living beings. Both βίος & βιός are spun into one double thread, a portmanteau βίός, a thread of life. It starts with "alive" and lines up to "death."

The thread of life exists in the cosmology of the antiquity. The three Fates, Moirai (literally, the "apportioners") are Clotho (spinner), Lachesis (allotter) and Atropos (unturnable, inflexible). Their task was to manage the life of all beings, in the form of a thread of life: to set the moment of their birth by the spinner, to ration the duration of their lives by the allotter, and to end the life by the unturnable cutting the thread. In the same way, the discourse and its subject are formulated from an unspun state of being. This point is Clotho, spinning the βιός out of the separate strands, known as zoë in biopolitics. Zoë is a co- and contra term for *bios* in antiquity. According to Giorgio Agamben (1998), *bios* stood for political life, a life in the society, the way of living proper, an ordered life, for oneself and in relation to others. In the current term, this is to live a socially sustainable life. Zoë then meant animal life, life excluded from the society, life of god(s): in other words, immortal life, life that cannot be killed, an abstract form of life. Therefore, it could be stated that zoë, in contrast to βιός, does not locate in time or as an acknowledged part of societal life. Still, βιός is made out of it by putting it into a linear form, as a bond between people. As the tributaries' course comes together and joins in the Kemi River, so do the different statements and fragments of speech, making a discourse along it. It is a foundation for a social order where the word order (n) is rooted in the Latin *ordinem*, a "row, line, rank; series, pattern, arrangement, routine," originally "a row of threads in a loom," and relating to Italic *ordiri*, "to begin to weave" (Online Etymology Dictionary, 2018h).

When the river becomes language, it is subjected to rationalizing by means and matters of measuring, classifying and standardizing. The presented resource and risk language are Lachesis, where the annual flood levels and hydropower productivity are calculated. This actualizes in the preparation of the flood maps that illustrate the number of people subjected to the risk of flood. The flow of waters is being apportioned and allotted: allotted as the hydropower companies calculate the overspill of water through their dams as their loss of profit. The three "Fates" comes from Latin *fata*, literally meaning "things spoken" from neuter past participle of *fari*, "to speak," as a divine prophesy for the predetermined course of life (Online Etymology Dictionary, 2018i). Therefore, the very word "fate" relies on the concept of speech, a discourse of risk that is woven with the stated forecasts and predictions.

If the acute accents are removed from the word, βιος means personal wealth and possession (Wiktionary, 2018c). In Marx's theory of value, the shaping up of the material to its final state brings out the value of what was already given in nature (Ingold, 2002, p. 80) as if the existing river were to become an energy resource. This pre-stated relation, in which man needs to process from the raw material, perfectly emphasizes the stage of the "unturnable" Atropos, where the "life" of the river is fulfilled in the moment it is cut and the flow ends, and this final form was enwoven with the very being of the river at the dawn of time. The "end" of the being is the fact that gives the "beginning" meaning, not only as the opposite, but also as the reference point, in between which everything else takes place.

Dis/cord over the river

The course of the river and the discourse along it are interwoven, which shows in the recent debate from the spring of 2018, not only in how the problem is understood, but also in how it is argued to be solved. The current level of flood protection in Rovaniemi along the Kemi River has been determined according to extremely rare flooding, estimated to emerge on average once in 250 years (Tennberg et al., 2018, p. 212). In Finland the general standard level of prevention is for a flood that emerges on average once in 150 years. The regional newspaper *Lapin Kansa* (Rytkönen, 2018, p. A5) argued that if the water level could not be controlled by planning measures, then the normative level of flood protection should be lowered as well. Such a statement would seem to have disastrous outcomes if the flood risk preparations were not made to meet the actual predicted flood levels but only to meet the politically appropriate level. In other words, to lower the standards to meet the level of preparation, not the actual level of flood. Flood only occurs or "becomes" when the line of actuality exceeds the line of prevention based on the potential risk.

In one of the newspaper articles, the question of the need to manage the Kemi River is entangled with its tributary, the Ounas River. The writer argues that the "protection of the Ounas River is a desertification scam" (Alatalo, 2018, p. O2). According to Urpo Alatalo, the legislation that is designed to protect the Ounas River as of 1983 has led to widespread desertification: there are no people or fish along and in the river. The river is "lifeless," but even during its "death stroke," the river still carries its hazardous nature with destructive floods. According to the writer, the only natural and efficient way is to capture the floodwaters in their birthplace. Here, the stage of nature or physically unregulated river (zoë) is considered as death, and controlling the source of the water is the resource for life.

The mayor of Rovaniemi, Esko Lotvonen, maintained that the decision of the Finnish government not to allow the water reservoir "has left the city at the mercy of natural forces" (Lapin Kansa, 2018, p. A1). In this course of discussion, nature is the end for the development of a community. An untamed river is a natural disaster, and a tamed river is a course to development. The river is to be considered as potentiality (potential threat or development potential), not as contingency. The issue will "rise again," as the mayor urged the appeal of the decision (Torvinen, 2018a, p. A5). An appeal was officially decided by the city on February 26, 2018 (Rovaniemi, 2018). The mayor of Pelkosenniemi municipality at the time, Pertti Severikangas, who opposed the reservoir project, predicted that the issue would reemerge within a decade and that the reemergence will be strongly colored by the approaching spring flood in Rovaniemi (Torvinen, 2018a, p. A5). As this statement strongly emphasizes, the decision on the reservoir is not seen as final; the finality manifests itself only when the potentiality has become actuality, and the reservoir discourse is woven directly into the course of the river in its materiality, flexing its administrative borders.

Because the city of Rovaniemi has meshed together flood management and hydropower projects, the city has not started constructing embankments, although embankments are required for a 50-year flood even if the reservoir were built (Torvinen, 2018b, p. A7). Such inactivity might suggest absentmindedness, but it could have another rationale behind it. Unpreparedness increases the risk of flood and thus also the tension on harnessing the waters of the river. The bow of Heraclitus is bent to a breaking point. Legal adviser Heikki Korpelainen from the Ministry of the Environment says that the reservoir is not the only option to solve the flood prevention objectives, and therefore there is no reason to weaken Natura 2000 protection (Torvinen, 2018a, p. A5). The claim is adequate if the flood prevention objectives are only seen as risk reduction, as a separate discourse from the regional development objectives to increase the volume of hydropower production.

Member of the Finnish Parliament and chair of the board of the Regional Council of Lapland Markus Lohi voiced his disappointment with the decision of the national government not to allow continuing with plans to build the water reservoir in the Kemi River branch. He nevertheless wanted to put an end, "a period," to promoting the water reservoir. The editorial states: "It is time to change the record" (Lapin Kansa, 2018, pp. A2, A4; Lohi, 2018, p. O2). These figures of speech emphasize the resource as something reoccurring, rising again, like the loop of the record, and how the coming of the reservoir follows a grammar. If the looping is equated to allotting, Lohi also pronounces the "unturnable" by firmly suggesting putting an end to the preparations. For Lohi, the inflexible is the Natura 2000 (Lohi, 2018, pp. O2–O3), a network of nature protection areas in the territory of the European Union, which is the death of the reservoir as they fight for the same geographical area. The big picture is interwoven from several threads of life, which by cutting are released from one another. The unborn reservoir would protect the city from the river, Natura 2000 protects the river from the reservoir and the river will either drown the Natura areas by becoming a reservoir or drown the city by running wild.

Lohi urges a switch from promoting the reservoir to promoting the development of the region by other means (Lohi, 2018, pp. O2–O3). This interestingly shifts the resource from natural to human resources. Because the exercise of social power requires the precondition of social consensus to maintain order, projects that are uncertain and meet a lot of societal resistance and risk the regional (rather than local) unity, the human resources for societal actions should be targeted elsewhere. In April 2019, the Finnish Association for Nature Conservation repeated the same argument while celebrating the decision of the Highest Administrative Court to put a full stop to the reservoir, as the "unturnable", and release its human resources (The Finnish Association for Nature Conservation, 2019).

Conclusions

The current debate on the river whirls around what may or will happen due to the changing climate and, based on these predictions, the necessity of proceeding with the management of the hydrological conditions of the river or infrastructural changes in city planning. Unpredictability needs to be turned into calculability. For governed society, the contingency emerges as an unpredictable risk, overflowing social life that can be managed only as potentiality. As Alt claims, in the discourse of (human) resource, "When the need for change becomes perceived as something that the environment necessitates, there is less need to govern the subjects per se" (Alt, 2016, p. 125). At this point, we are talking not merely about the natural powers as a resource, but the question has turned to the people as a resource, as a pairing element for the nature, as the warp and the weft constituting each other into a fabric.

If the phenomenon becomes communicative, it can be turned into a resource for social order, a line on a paper or a vector on a map in this ordered, grammatical and rationed world. In the case of Rovaniemi, it can be argued that the process triggered by the middle-European urban areas is implemented in northern, much more rural areas when it comes to the density and scale of their infrastructure, and this discourse is enforced by flood-risky planning to drive certain adaptive measures while ignoring others or by scaling them out by raising the level of risk. The bow of Heraclitus loses its shape, if the βίός bending the bow does not manifest itself as the thread of life and death, possibility and risk.

Resource extraction commonly transcends the community that spreads around it. In the current discourse of the river, it seems that if the resource is not managed, it will wash away the community. Historically, inhabitation was born around the river because the river enabled mobility, but while the dwellings along it turned into static structures, the living requirements seem to demand the immobilization of the river's dynamics. The river has become not only a risk for the community, but also people's personal risk due to changes in flood damage compensation legislation. People's potentiality to become, whether to become residents, labor, beneficiaries or secured, necessitates that the river evolve from what it may be to what it will be and, therefore, what it must be. Pull a thread, and the fabric will fall apart. Pull a threat, and the social order tightens. Spill water on it, and the water felts the separated life threads together into one solid piece.

References

Agamben, G. (1998) *Homo sacer: sovereign power and bare life*. Stanford: Stanford University Press.

Alatalo, U. (2018) "Lappi puhuu: Ounasjoen suojeleminen oli autioittamistemppu," *Lapin Kansa*, May 16, p. O2.

Alt, S. (2016) *Beyond the biopolitics of development: being, politics and worlds*. Rovaniemi: Lapland University Press.

Anderson, V. and Johnson, L. (1997) *Systems thinking basics: from concepts to causal loops*. Waltham: Pegasus Communications.

ELY Centre Lapland (2003) *Ounasjoki* [Online]. Available at: www.ymparisto.fi/fi-FI/Luonto/Suojelualueet/Natura_2000_alueet/Ounasjoki(6146) (Accessed: February 15, 2019).

The Finnish Association for Nature Conservation (2019) *Vuotoksen tekoaltaalle lopullinen hylkäys oikeudesta* [Online]. Available at: www.sll.fi/2019/04/11/vuotoksen-tekoaltaalle-lopullinen-hylkays-oikeudesta/?fbclid=IwAR1lb1cHiLEOspoJXHIXZl-2bS93Zk7fYR71x2SBHqmJpYlKQyy9zvFyiqI (Accessed: April 11, 2019).

Foucault, M. (1977) "Nietzsche, genealogy, history" in Bouchard, D.F. (ed.) *Language, counter-memory, practice: selected essays and interviews*. Ithaca: Cornell University Press, pp. 139–164.

Haugaard, M. (2003) "Reflections on seven ways of creating power," *European Journal of Social Theory*, 6(1), pp. 87–113.

Ingold, T. (2002) *The perception of the environment: essays on livelihood, dwelling and skill*. London: Routledge.

Kangas, R. (2009) "The market, values and coordination of actions: value integration to libertas indifferentiae," *Journal of Classical Sociology*, 9(3), pp. 291–318.

Kemijoki Ltd. (2018) *Voimalaitokset ja tuotanto* [Online]. Available at: www.kemijoki.fi/toimintamme/voimalaitokset-ja-tuotanto.html (Accessed: February 15, 2019).

Keskitalo, E.C.H. (2013) "Introduction. Local organisation to address flood risks: possibilities for adaptation to climate change?" in Keskitalo, E.C.H. (ed.) *Adaptation and extreme events at the local level*. Cheltenham: Edward Elgar Press, pp. 1–34.

Keskitalo, E.C.H., Åkermark, J. and Vola, J. (2013) "Flood risks along the Torne River between Sweden and Finland" in Keskitalo, E.C.H. (ed.) *Adaptation and extreme events at the local level*. Cheltenham: Edward Elgar Press, pp. 67–94.

KHO Supreme Administrative Court (2017) *KHO:2017:87* [Online]. Available at: www.kho.fi/fi/index/paatoksia/vuosikirjapaatokset/vuosikirjapaatos/1494937158863.html (Accessed: February 15, 2019).

Kivelä, K. (2016) "Vuotoksen allas on jokiyhtiön hanke. Hyöty verotuloina jää todella vähäiseksi aiheutettuun haittaan nähden," *Maaseudun tulevaisuus*, December 21 [online]. Available at: www.maaseuduntulevaisuus.fi/mielipiteet/lukijalta/vuotoksen-allas-on-jokiyhti%C3%B6n-hanke-1.173183 (Accessed: February 15, 2019).

Laitala, M. (2013) "Vuotoksen allas haudattiin lopullisesti: suot luonnonsuojelualueeksi," *Tekniikka & Talous*, August 14 [Online]. Available at: www.tekniikkatalous.fi/tekniikka/energia/2013-08-14/Vuotoksen-allas-haudattiin-lopullisesti-suot-luonnonsuojelualueeksi-3314844.html (Accessed: February 15, 2019).

Lakoff, G. and Johnsen, M. (2003) *Metaphors we live by*. London: The University of Chicago Press.

Lapin Kansa (2018) "Pääkirjoitus/On aika vaihtaa levyä," *Lapin Kansa* (editorial), February 2, p. A2.

Laukkanen, M. (2013) "Kymmenien miljoonien menetykset vesivoiman ohijuoksutuksissa," *STT info*, May 2 [Online]. Available at: www.sttinfo.fi/tiedote/kymmenien-miljoonien-menetykset-vesivoiman-ohijuoksutuksissa?publisherId=4627873&releaseId=4899264 (Accessed: February 15, 2019).

Lohi, M. (2018) "Kemihaaran allasta ei rakenneta," *Lapin Kansa*, February 2, pp. O2–O3.

Matthew 7:13–42, Holy Bible: New International Version.

Online Etymology Dictionary (2018a) *Resource (n.)* [Online]. Available at: www.etymonline.com/word/resource#etymonline_v_12902 (Accessed: February 15, 2019).

Online Etymology Dictionary (2018b) *Source (n.)* [Online]. Available at: www.etymonline.com/word/source#etymonline_v_23927 (Accessed: February 15, 2019).

Online Etymology Dictionary (2018c) *Power (n.)* [Online]. Available at: www.etymonline.com/word/power (Accessed: February 15, 2019).

Online Etymology Dictionary (2018d) *Course (n.)* [Online]. Available at: www.etymonline.com/word/course#etymonline_v_46719 (Accessed: February 15, 2019).

Online Etymology Dictionary (2018e) *Discourse (n.)* [Online]. Available at: www.etymonline.com/word/discourse (Accessed: February 15, 2019).

Online Etymology Dictionary (2018f) *Reservoir (n.)* [Online]. Available at: www.etymonline.com/word/reservoir (Accessed: February 15, 2019).

Online Etymology Dictionary (2018g) *Economy (n.)* [Online]. Available at: www.etymonline.com/word/economy (Accessed: February 15, 2019).

Online Etymology Dictionary (2018h) *Order (n.)* [Online]. Available at: www.etymonline.com/word/order (Accessed: February 15, 2019).

Online Etymology Dictionary (2018i) *Fate (n.)* [Online]. Available at: www.etymonline.com/word/fate#etymonline_v_1153 (Accessed: February 15, 2019).

Przybyslawski, A. (2003) "The bow of Heraclitus: a reflection on the languages of becoming" in Tymieniecka, A-T. (ed.) *The passions of the soul in the metamorphosis of becoming. Islamic philosophy and occidental phenomenology in dialogue*, vol. 1. Dordrecht: Springer, pp. 155–160.

Rovaniemi (2018) *Kaupunginhallitus käsitteli: tilahallinnan järjestäminen, valitus tulva-allasasiassa, palotoimen ja ensihoidon tilojen hankesuunnittelu sekä alueteatterin yhtiöittäminen* [Online]. Available at: www.rovaniemi.fi/news/Kaupunginhallitus-kasitteli-tilahallinnan-jarjestaminen,-valitus-tulva-allasasiassa,-palotoimen-ja-ensihoidon-tilojen-hankesuunnittelu-seka-alueteatterin-yhtioittaminen/26871/183f25ed-f7ff-4b39–9ce7–701a9780fc33 (Accessed: February 15, 2019).

Ruuska, T. (2017) "Capitalism and the absolute contradiction in the Anthropocene" in Heikkurinen, P. (ed.) *Sustainability and peaceful coexistence for the Anthropocene*. New York: Routledge, pp. 51–67.

Rytkönen, P. (2018) "Uutiset/Tulva-allasta ei tule Kemihaaraan," *Lapin Kansa*, February 2, pp. A4–A5.

Talvitie, M. (2016) "Vesi satoi Kemijoki Oy:n laariin – tuotanto kasvoi kesällä viidenneksen," *YLE Uutiset* (Economy), August 18 [Online]. Available at: https://yle.fi/uutiset/3-9103639 (Accessed: February 15, 2019).

Tennberg, M., Vuojala-Magga, T., Vola, J., Sinevaara-Niskanen, H. and Turunen, M. (2018) "Negotiating risk and responsibility: political economy of flood protection management in Northern Finland" in Hiyama T. and Takakura H. (eds.) *Global warming and human-nature dimension in Northern Eurasia*. Singapore: Springer, pp. 207–221.

Torvinen, O. (2018a) "Allas kaatui, kun ei ole ainoa tulvasuojelukeino/Esko Lotvonen tuohtunut: 'Rovaniemen pitää valittaa'/'Kymmenen vuotta ehkä voidaan olla rauhassa'," *Lapin Kansa*, February 2, p. A5.

Torvinen, O. (2018b) "Uutiset/Vallit etenevät, paitsi Rovaniemellä," *Lapin Kansa*, February 21, pp. A6–A7.

van Ree, C.C.D.F., Willemsen, M.A.M., Heilemann, K., Morris, M.W., Royet, P. and Zevenbergen, C. (2011) "FloodProBE: technologies for improved safety of the built environment in relation to flood events," *Environmental Science & Policy*, 14(7), pp. 874–883.

Wiktionary (2018a) *βίος* [Online]. Available at: https://en.wiktionary.org/wiki/βίος (Accessed: February 15, 2019).

Wiktionary (2018b) *βιός*. [Online]. Available at: https://en.wiktionary.org/wiki/%CE% B2%CE%AF%CE%BF%CF%82 (Accessed: February 15, 2019).

Wiktionary (2018c) *βιος* [Online]. Available at: https://en.wiktionary.org/wiki/%CE% B2%CE%B9%CE%BF%CF%82 (Accessed: February 15, 2019).

YLE (2011) *Vuotoksen allashankkeen vaiheita* [Online]. Available at: https://yle.fi/aihe/ artikkeli/2007/08/20/vuotoksen-allashankkeen-vaiheita (Accessed: February 15, 2019).

5 "Prudent development"

The (r)evolution of the Arctic energy concern in the 2007–2017 Arctic Energy Summit Reports

Hanna Lempinen

Introduction: the Arctic as the world's new energy province?

In popular, political and scholarly debates alike, the Arctic region is seen as about to become the world's new energy province. A significant share of the world's unutilized oil and gas resources are estimated to be located in the Arctic (USGS, 2009). The international interest in the region's energy endowments has heightened in the interplay of various overlapping and interconnected developments. The expected growth of global energy demand (IEA, 2017), dwindling reserves at existing production sites (cf. Owen et al., 2010; Di Muzio and Salah Ovadia, 2016) and anxiety over the impacts of political events on energy supplies (cf. e.g. Liuhto, 2009; Paillard, 2010) all play an important role, as does climate change. Together with technological developments, the thawing sea ice is expected to make previously inaccessible areas available for energy production and transportation activities (cf. e.g. Loe and Kelman, 2016, p. 25; Kristoferssen, 2014, p. 56).

While this "widely circulated, orthodox version" (Hannigan, 2015, p. 8) of what energy means in the Arctic – or, conversely, what the Arctic means in the context of global energy – has gained a significant foothold, it has also been brought to question on many fronts. For one, it is debatable whether the estimated reserves actually exist and, if they do, whether they can be extracted in a manner that is both economically profitable and compatible with international climate commitments and goals (McGlade and Ekins, 2015). Additionally, the warming climate might not only reduce the ice cover but also lead to more extreme and more unpredictable weather conditions, making energy operations in the North much riskier both operationally and financially (Emmerson and Lahn, 2012; Harsem et al., 2011). Woven together, these arguments construct a rather different kind of Arctic region, one that is and will remain "more of an energy backyard than a frontier" (Sidortsov, 2016, p. 2).

While the energy narrative laid out here is based on a simplistic understanding of energy as the exports of oil and gas (cf. Lempinen, 2018), it is also problematic in that it focuses on the role that the energy resources of the Arctic region have in the global puzzle of energy production and energy security. Meanwhile, only marginal attention is paid to what the energy concern entails in the context of the Arctic region or what it means for the residents and communities in the circumpolar North (see also Sidortsov and Sovacool, 2015). These questions are also at the core of discussions related to the social dimension of the Arctic energy concern and the social sustainability of Arctic energy developments. In sustainability-related debates in general, the social dimension has been broadly acknowledged as elusive and nearly impossible to operationalize (Boström, 2012; Murphy, 2012). This remains even more the case in the specific context of the energy, which has predominantly been framed as a techno-econo-scientific concern, with little attention paid to what its societal dimensions beyond socioeconomic development might entail (see Lempinen, 2017).

To make up for this deficiency, I will focus explicitly on the ways in which the regional energy concern has been conceptualized from within and for the Arctic region. Making use of the final reports from the 2007–2017 Arctic Energy Summits, a series of high-level science and policy meetings with an aim to highlight the regional dimensions of Arctic energy, the chapter scrutinizes the regional energy concern from within the region. It explores how the needs, challenges and special features of the northern energy concern have been conceptualized in the Arctic during the last decade. In addition, the focus of the analysis lies in the ways in which the social dimension of energy in the Arctic region is framed and portrayed and in the ways in which the present and the future of regional energy development are entwined with and constructed as a social sustainability concern.

The Arctic Energy Summits: an overview

The Arctic Energy Summits – an outgrowth of the International Polar Year 2006–2008 initiative #299 led by the Institute of the North based in Anchorage, Alaska, and later "adopted" by the Arctic Council's Sustainable Development Working Group – are a series of high-level science and policy meetings dealing with issues related to energy in the context of the circumpolar North. The summits draw "several hundred industry officials, scientists, academics, policymakers, energy professionals and community leaders together to collaborate and share leading approaches on Arctic energy issues" (Arctic Energy Summit, 2018a). The first summit took place in Anchorage, Alaska, in 2007; the second in Akureyri, Iceland, in 2013; the third in Fairbanks, Alaska, in 2015; and, at the time of this writing, the most recent one was organized in Helsinki, Finland, in 2017.

What makes the summits different from the wealth of conferences touching upon or explicitly dealing with Arctic energy is their focus on energy as a regional concern as well as the emphasis on the diversity of the energy concern within the region. In contrast with the "exploit- and export-" oriented mainstream approach to energy in the Arctic region, under the summits, energy resources are (also) discussed "from a comprehensive and holistic perspective" and as a "fundamental building block for communities and economies" with an aim to develop best practices, minimize risks and "foster greater benefits to northern peoples" (Arctic Energy Summit, 2018b). The findings and conclusions of each of the summits have been published in dedicated final reports that together cover 10 years of regional debates over energy in the Arctic region. As a result, they provide a possibility for exploring the ways in which the regional understanding of energy has changed and evolved during the period of scrutiny, which saw both a growing interest in Arctic energy resources and an increasing awareness of the climate impacts of (Arctic) energy development (cf. McGlade and Ekins, 2015). While the reports are at best superficial and simplistic summaries that inevitably flatten the diversity of voices and viewpoints present(ed) at the summits and their sessions, their analysis is nonetheless both important and informative. The final report documents are intended to provide the knowledge base that will support the Arctic Council's Sustainable Development Working Group in its future work touching upon energy and communities in the circumpolar North.

Methodologically, this chapter scrutinizes the summits' final reports through a loose framework of thematic content analysis (cf. e.g. Julien, 2008; Pickering, 2004): the contents and constituents of the regional energy concern as well as their change over the decade between the first and the fourth, most recent summit, are mapped in a data-based, qualitative manner. In the analysis, the focus is on the sections that aim to summarize the contents, key concerns and recommendations from each of the conferences, but the abstracts of individual conference presentations listed at the end of some reports have been left outside the scope of scrutiny. The chapter focuses on how Arctic energy is framed and how its intertwinements with societal life and social sustainability in the North are constructed, paying special attention to how the uneasy relationship between sustainable development and nonrenewable resource extraction keeps being constructed, negotiated and justified.

The global significance of Arctic energy wealth vs. regional energy security needs

Despite the regional focus of the Arctic Energy Summits, the final reports begin with and echo the ways in which the Arctic region and its energy reserves are framed in international political and media representations. References to the energy "richness" (AES, 2013, cover) of the region are many. The Arctic is portrayed as "abundant" (AES, 2015, p. 5) in its oil and gas reserves, housing deposits "possibly greater than 25 percent of global reserves" (AES, 2010, p. 26) left for extraction. These estimated massive reserves frame "the High North as a source of new energy for the world" (AES, 2010, p. 6), even to the extent that developing the Arctic as "an energy province is essential to global and energy security" (AES, 2010, p. 22). The region's energy reserves are allotted a pivotal role in fueling global economic growth (AES, 2013, p. 21), as well as supplying "petroleum products like plactics, chemicals, and a vast array of synthetic materials that originate from petroleum" that contemporary societies are "highly dependent on" (AES, 2017, p. 9).

However, the reports don't just deal with energy as an issue of feeding global energy and materials demand and economic growth; energy is also constructed as an issue of regional energy challenges and needs. Both of these are factors more often than not left out of global framings of energy in the Arctic region; there is therefore "a need to assess local need (access and affordability) versus the global needs" (AES, 2013, p. 29). Indeed, the energy summits and their final reports portray the Arctic region as one of both "great energy wealth" and "great energy poverty" (AES, 2010, p. 13; also AES, 2017, p. 7). Amid the "richness" (AES, 2013, cover), the residents of the vast region are suffering from "crippling energy costs threatening their very existence" (AES, 2010, p. 19) and are paying "some of the highest energy prices in the world" (AES, 2015, p. 5). In the shadow of the world's new energy province, the regional energy concern revolves around the double challenge of "security and affordability" (AES, 2015, cover): access to affordable (AES, 2010, p. 9) and reliable (AES, 2010, p. 24) energy resources "dictates" (AES, 2010, p. 8) the future of rural subsistence lifestyles (AES, 2010, p. 11), "quality of life" (AES, 2010, p. 8) and the very survival (AES, 2010, p. 9) of northern communities. From this perspective, the availability and affordability of energy in the circumpolar North become pivotal social sustainability concerns.

These concerns repeatedly raised throughout the reports are not unfounded as the cold climate, long distances and heavy industries make the Arctic nations among the highest per capita energy consumers in the world (Rasmussen and Roto, 2011, p. 151). Thus, regional energy development is "of significant interest to the energy and security needs of the Arctic nations" (AES, 2010, p. 8). Solutions for the stabilization of energy costs and measures promoting energy self-sufficiency in northern remote communities thus occupy a key position in discussions of the state and future of energy in the circumpolar North (AES, 2013, p. 18).

However, it is not the oil and gas resources of the region but rather the renewable energy reserves that are brought to the fore. Indeed, the Arctic is portrayed as "abundant" not only in its hydrocarbon reserves, but also in its renewable energy endowments: "wind, solar, hydrokinetic and geothermal" (AES, 2015, p. 5). Development and utilization of renewable energy resources are seen to bring "opportunities . . . in the form of generating lower costs, while also improving the energy self-sufficiency of communities" (AES, 2013, p. 15). This will both efficiently address the dual energy security challenge of affordability and availability (for energy security, see e.g. Lempinen and Cambou, 2018; Kruyt et al., 2009; Cherp and Jewell, 2014) and contribute to the "potential for more success for communities and countries" in the North (AES, 2017, p. 6). However, the prospects of renewable energy deployment in the region are hampered by lack of knowledge as the decades of political, economic and scholarly attention on the nonrenewable resources of the region have led to a situation where there "hasn't been enough effort to examine renewable and alternative energy" (AES, 2013, p. 29). The reports' references to the dire need to map the renewable energy potential of the region are, however, making an impact. The region's renewable energy potentials are in the process of being mapped under the umbrella of the Arctic Renewable Energy Atlas (AREA, 2018), a project endorsed by the Arctic Council's Sustainable Development Working Group.

Alongside renewable energy sources, new and emerging energy sources and technologies also play a crucial role in responding to the region's energy needs. Technological development is seen as a way of lowering the costs of electricity, heating and transportation (AES, 2010, p. 17), as well as improving the reliability of energy supply. Diversification of power generation systems (AES, 2015, pp. 5, 7), improving grid access (AES, 2013, p. 5), retrofitting existing diesel-fueled energy systems (AES, 2010, p. 11; 2017, p. 3) and developing technologies and incentives for energy efficiency and saving (e.g. AES, 2010, p. 11; 2013, p. 21, 2015, p. 6, 2017, p. 10) all contribute to improved energy security and "economic opportunity" (AES, 2010, p. 15). In addition, technological advancements are expected to bring completely new energy sources to the local Arctic energy mixes: different forms of "emerging energy technology" (AES, 2013, p. 10) ranging from hydrogen to tidal and ocean heat utilization and even to floating nuclear power plants (AES, 2013, pp. 9–10, 15) are being experimented with. In sum, such "a wide array of energy production, storage, and transfer technology is under development" (AES, 2015, p. 5) that "[t]echnology in general is not the problem. Technical solutions exist for each challenge area" (AES, 2010, p. 10). The problems of regional energy poverty and energy security could already be solved.

Fueling development

While catering to regional energy needs constitutes a major component of the Arctic Energy Summits' concerns that deal with the societal aspects of (energy) sustainability, Arctic energy as it is portrayed in the reports cannot be reduced to an issue of improving energy security in the circumpolar North. It is also the instrumental role that energy development is expected to have in fueling "local and regional development" (AES, 2013, p. 25) and "sustainable development" (AES, 2015, p. 5) that receives the bulk of the attention in the reports. Indeed, energy resources and their secured availability are not only a "fundamental component of sustainable development" and "human development" (AES, 2013, p. 3, 2015, p. 5) and a "fundamental building block for communities" (AES, 2015, p. 3) but also vital for regional economic activity and "the economic development needed to retain a strong and vibrant population" (AES, 2013, p. 6). In particular, extractive energy development is framed in terms of the "large economic opportunities" it can bring (AES, 2013, p. 21) and as vital in "the creation of wealth for governments, communities and the private sector" (AES, 2015, p. 5) as it contributes to "economic and community prosperity" (AES, 2013, p. 4) and "produces revenue at all three levels of government, as well as important jobs for citizens" (AES, 2013, p. 22). These remarks highlight the second dimension through which Arctic energy becomes a (social) sustainability concern: the revenues derived from energy-related activities play an instrumental role in fueling and supporting overall regional development.

Indeed, "responsible development and utilization of Arctic energy resources" are seen to "have great potential to spur community and economic resilience" (AES, 2013, p. 3). However, alongside the great economic opportunities and bright development prospects that making use of regional hydrocarbon reserves is expected to bring, the problems and challenges associated with the development of Arctic energy are also discussed at great length. Consequently, regional energy development cannot be framed solely as an opportunity and a strategy for regional sustainable development. It is also a sustainability challenge whose solution the reports address in great detail. The final reports do not deal with the question of *whether* to go forward with developing the Arctic as the world's new energy province as much as they engage in discussions of and criteria for *how* this development should be conducted. In fact, it is seen as *"necessary to depoliticize the question of whether the development of energy resources in the Arctic can be done in a sustainable and responsible way"* (AES, 2013, p. 8; emphasis added). While "there will always be an element of risk in resource development" (AES, 2013, p. 8), the "enormous positive externalities" (AES, 2013, p. 8) that regional energy development is expected to bring justify taking these environmental and societal risks as long as they are managed and mitigated according to best practices and the most stringent of standards possible (see AES, 2013, pp. 14–15; 2015, pp. 10–11).

Descriptions and definitions of how to correctly implement energy development in the Arctic thus fill the pages of the final reports: according to these reports, Arctic energy projects "should seriously consider environmental impacts and should adopt current best practices to minimize harm while also taking into account regional, national and global energy needs" (AES, 2010, p. 16); be "balanced with protection of the environment and respect for traditional ways of living" (AES, 2013, p. 3, 2015, p. 5); balance "risk mitigation, cultural integrity, and economic opportunity" (AES, 2015, p. 5; also AES, 2013, p. 3) and address the "three key topics" of the "economics, environment and impact on the people of the North" (AES, 2010, p. 16). Through repeated references to the separate dimensions of the environment, the economy and the social dimension in the North, the reports echo and reinforce the definitions of sustainability articulated in the Brundtland Commission's report (see WCED, 1987). However, in addition to and instead of sustainability (AES, 2010, pp. 9, 17, 2017, p. 7), the requirements of Arctic energy developments are framed through terms such as "responsibility" and developing "resources *prudently*" (AES, 2013, p. 3; 2015, p. 5; emphasis added). According to the 2015 Summit's Final Report, this new term "encompasses a holistic goal including economic, environmental and social development" (AES, 2015, p. 12). While the contents of this concept are far from novel – being basically identical to those familiar from the sustainability debates – resorting to the term "prudent" instead of "sustainable" avoids much of the critical discussion that revolves around the idea of sustainably developing (with) nonrenewable resource extraction and making use of finite resources to create long-lasting benefits and wealth (see Lempinen, 2018).

In the reports' definitions of "prudent" development, attention is devoted to both the "extremely sensitive" Arctic environment (e.g. AES, 2010, p. 8; also AES, 2013, p. 4; 2015, p. 5) and the "needs of local peoples, communities and economies" (AES, 2015, p. 6). Out of these, especially the latter deserves closer scrutiny, partly because of the ways in which the elusive social dimension of sustainability has in general been sidelined from sustainability-related debates equally in the context of energy and beyond (cf. Lempinen, 2018) and partly owing to the ways in which the social or societal dimension is framed in the context of the Arctic Energy Summits and their final reports. The impacts of energy-related developments, whether positive or negative, "have historically been unevenly distributed" (AES, 2010, p. 23), and it has been questioned whether "local communities receive enough benefits to offset downstream environmental and social costs" (AES, 2015, p. 11). Against this background, in their "regional approach to energy" (AES, 2015, p. 5), the reports emphasize the "immediate, local benefit" (AES, 2015, p. 6) experienced by "the People of the North" (AES, 2013, p. 11), "all northern residents (AES, 2015, p. 5) and "specifically those indigenous people living a subsistence lifestyle in remote communities" (AES, 2010, p. 23) and "traditional or subsistence-based economies" (AES, 2015, p. 6).

In this, the summits on the one hand take a regionalized approach to the social dimension, paying only marginal attention to "global implications of activity in the region" (AES, 2015, p. 6). On the other hand, they adopt a strongly developmental approach to what social sustainability in the context of Arctic energy entails, focusing on the concrete benefits gained by northern communities as a direct consequence of oil and gas development (see Vallance et al., 2011). However, the social dimension is not wholly reduced to developmental terms; issues related to local and indigenous participation and consultation (e.g. AES, 2013, pp. 4, 13; 2015, p. 7; 2017, p. 7) and integration of traditional knowledge (e.g. AES, 2010, p. 23; 2013, p. 17; 2015, p. 11; 2017, p. 8) into energy-related decision-making also feature prominently when the social requirements of energy developments are discussed. The reports frame "local communities and indigenous peoples as key stakeholders, rights-holders and energy partners" (AES, 2017, p. 7), and for this reason, "the utilization of Arctic resources will not occur without coming into conflict and subsequent dialogue with *the indigenous people who call the Arctic home*" (AES, 2013, p. 4; emphasis added). It is "important that local peoples are at the table and not represented or misrepresented by outside interests (AES, 2017, p. 7) and that projects "move forward very carefully and collaboratively" (AES, 2010, p. 24). The benefits of this include "improved project planning and design, better decision making, and ensuring more equitable benefits to community and Indigenous stakeholders" (AES, 2017, p. 7).

In emphasizing consultation, communication and participation, the summits' approach to what socially sustainable Arctic energy development entails also borrows from procedural understandings of sustainability. From this perspective, both the concrete outcomes of "development" and the ways in which this development is experienced by those affected need to be taken into account and addressed (see Del Río and Burguillo, 2008; Vanclay, 2002, 2003). What must be noted, however, is that the participatory aspects of Arctic energy development are also framed instrumentally. Informing and engaging the residents of the Arctic region "have a role in strengthening public confidence as well as spreading awareness of change and activity increases" (AES, 2013, p. 4); "stakeholder engagement" (AES, 2013, p. 12) is seen as a means to "ensure local support for Arctic energy projects" (AES, 2010, p. 24) and to "promote the public's ownership in the outcome" (AES, 2013, p. 17) for "maintaining public confidence" (AES, 2013, p. 5). As a result, engagement of and collaboration with local residents is not (solely?)

a value in its own right. Developers "utilize emotion, traditional values, scientific evidence, lessons learned and lessons yet to be learned, to build relationships" (AES, 2017, p. 8) to maximize the support from local communities and, through this, to improve the conditions of their own energy development projects. All in all, the discussion on engagement takes place against an uneasy backdrop. While the most recent report explicitly states that "engagement at the local level must include the option to say 'yes or no' to development" (AES, 2017, p. 7), the "unequal relationship between local community and industry" is also an acknowledged fact: "a small community will rarely have the financial resources and capacity of resource development groups" (AES, 2013, p. 13).

The elephant in the living room? Climate change and Arctic hydrocarbon development

From the perspective of the social dimension in the North, the Arctic Energy Summits and their final reports revolve around two key concerns. On the one hand, the discussion focuses on aspects of energy availability and affordability, on eliminating Arctic energy poverty and catering to regional energy needs predominantly through renewable energy resources and technological development. On the other hand, energy and the social dimension in the Arctic region converge in the context of extracting and exporting fossil fuels to create wealth and well-being for the socioeconomically challenged region. Oil and gas developments are framed as vital in terms of sustainable and long-lasting development of Arctic societies, but what these interpretations and expectations do not account for are the climate impacts of Arctic oil and gas developments. In other scientific assessments and reports produced under the auspices of the Arctic Council, climate change has repeatedly and consistently been pointed out as one of the direst challenges (already) facing Arctic societies and communities (see AHDR, 2015; ARR, 2016). At the same time, it is well established that the energy sector is responsible for around two thirds of the annual global greenhouse gas emissions (IEA, 2015, p. 11); furthermore, the year 2017 saw record-high carbon dioxide emissions both from the energy sector and overall (IEA, 2018, p. 3).

Against this background, the implications of climate change on the prospects of Arctic energy development appear challenging from at least two perspectives. Adding even a "modest Arctic energy boom" (Forbis and Hayhoe, 2018, p. 2) to the global emissions balance would send the world far beyond emission levels deemed acceptable or safe under the two-degree global warming target set in the Paris Agreement (UNFCCC, 2016). Such an increase in emissions would only accelerate the impacts of climate change already felt in the Arctic region. Relying on fossil fuels as a developmental strategy when global emissions will need to be limited is risky also in the sense that the future demand for oil and gas will most likely not be aimed at the Arctic reserves that have been estimated as notoriously risky and expensive to utilize (Emmerson and Lahn, 2012). Staying within the two-degree global warming target has been calculated to mean utilizing only a fraction of the currently known resources that are located elsewhere in the world and leaving practically all capital-intensive Arctic hydrocarbon reserves unutilized (see McGlade and Ekins, 2015). Indeed, in the context of fossil fuel extraction, the activity and their revenues do not necessarily – or even likely – last to the point when the existing reserves have been fully utilized. Instead, their utilization ceases at the point when their extraction is no longer economically profitable, politically supported or societally accepted (Freudenburg, 1992, p. 324; Mitchell et al., 2001).

Despite their lasting focus on hydrocarbon-based regional development, the summits' final reports and proceedings also echo the issue of climate change. This is reflected in the remarks referring to the need to curb the climate gas emissions of Arctic oil and gas activities. Repeated references are made to technologies and policies encouraging carbon capture and storage in energy production (AES, 2010, pp. 11, 15, 17; 2017, p. 10) and, more recently, also to carbon pricing and emission trading schemes (AES, 2017, p. 10). What is of course noteworthy in these references is their focus on upstream emissions: all in all, minimal attention is paid to what happens when the resources produced in the Arctic – even with zero greenhouse gas emissions – are consumed elsewhere in the world. The challenges posed by climate change are also pondered in the context of the impacts that the changing climate will have on Arctic energy activities themselves. While "climate change and its impact on the Arctic are well documented, how this will play out in the development of energy projects is less clear" (AES, 2010, p. 16) as are the impacts of climate change on Arctic (energy) infrastructure (AES, 2010, p. 20). The need to "integrate climate change into existing management" (AES, 2015, p. 6) is acknowledged, although the practical implications of this need are not addressed in any concrete manner.

What must be noted is that while the summits' foci on regional energy needs and criteria for sustainable Arctic energy development have remained somewhat consistent and unchanged during the decade and the four summits, it is in the reports' relationship with energy production and climate change that we can see a dramatic discursive change. Where the first summit report still refers to the "need to increase, not curtail fossil fuel production" as a bridge to future alternative technologies (AES, 2010, p. 15) and to "a paradigm shift that will allow for the world to view coal as a transformational fuel and a transitional hydrocarbon resource, rather than a "dirty" combustion fuel" (AES, 2010, p. 11), the 2017 summit – arranged after the Paris Agreement (see AES, 2017, p. 5) – describes climate change as "a *new* framework" (AES, 2017, p. 5; emphasis added) for addressing Arctic energy issues and the future prospects of Arctic oil and gas development as "clearly a flashpoint for debate" (AES, 2017, p. 5). The 2017 report refers to "commitments made in the Paris Agreement" (AES, 2017, p. 5), an ongoing energy transition toward renewables and alternative energy sources (AES, 2017, pp. 9, 11) and a "low-carbon" (AES, 2017, p. 9) and "carbon-neutral" (AES, 2017, p. 11) future. These new emphases are clearly consistent with the need to increase the affordability and availability of energy in the region through renewable energy and new technologies, but what remains less clear is how they will play out in the context of filling the socioeconomic void left by reducing and potentially ceasing Arctic oil and gas extraction activities.

Concluding thoughts

This chapter has closely read the final reports of a decade of Arctic Energy Summits, a series of high-level science and policy meetings dealing with issues related to energy in the context of the circumpolar North, with an aim to investigate how the Arctic energy concern and its social sustainability dimension have been conceptualized from within the region. This focus has been motivated by at least two reasons. On the one hand, the social dimension (of sustainability) as it pertains to energy has been widely acknowledged as undertheorized and understudied; on the other hand, the regional aspects and importance of energy in the Arctic region have consistently been overshadowed by the international interest in the role that the region's oil and gas reserves are expected to have in fueling the world's increasing energy demand.

While the energy summits and their final reports acknowledge and underline the importance of the Arctic region for the world's future energy supply and construct the region predominantly in terms of its massive energy wealth, the implications of this wealth are constructed in a different light. Arctic energy reserves – especially the region's massive energy resource endowments – play a vital role in supplying for the demand within the region whose residents continue to "face affordability, reliability and security concerns" (AES, 2017, p. 5). And yet, the oil and gas resources of the region are or at least have been reserved a crucial role in creating revenues and wealth that will have an instrumental role in the future development of the region and its communities and societies. While the concerns over regional energy security and the idea of using energy to fuel overall (sustainable?) development have consistently been foregrounded by the summits in the last ten years, the increasing awareness of and worry over climate change and its impacts have contributed to a gradual but remarkable shift in the reports' relationship to Arctic hydrocarbon development. Previously advocated rather unreservedly, provided that carbon dioxide emissions arising from production processes are minimized (AES, 2010), fossil fuel extraction is, in the most recent report, described in terms of a "flashpoint for debate" (AES, 2017, p. 5), and the overall discussions "regarding the future of the Arctic" are framed as "dominated" by discussions of "innovation and renewable energy" (AES, 2017, p. 9).

While this shift from fossil fuels to renewables and to a "green energy transition" (AES, 2017, p. 5) has most likely taken place at the level of discourse rather than actual energy policy and practice, it is nonetheless notable. What is equally notable is the way in which the Arctic continues to be framed as the "new energy province" (AES, 2010) for the world but in a different context than before. The first summit report emphasized the impact that the Arctic oil and gas resources have on "how the world defines energy security" (AES, 2010, p. 24), but during the decade that the summits have been arranged, the importance of Arctic energy for the world has increasingly become constructed in terms of the "potential" that the region has for being "a model of clean energy development and use" (AES, 2013, p. 4). This shift is at its clearest in the most recent, the 2017 summit report, which outlines a future where "Arctic energy expertise can be exported to help other regions of the world adapt or transition to cleaner energy" (AES, 2017, p. 5) and where the "Arctic is increasingly defined by its renewable energy and energy efficiency leadership" (AES, 2017, p. 5). Thus, despite the change in the content of what "energy" in the Arctic will or is expected to entail, what has not changed are the envisioned significance and exceptionally weighty role that continue to be reserved for Arctic energy in solving both the energy challenges of the world and the sustainable development challenges of the Arctic region.

References

AHDR (2015) *Arctic Human Development Report II: regional processes and global linkages*. Akureyri: Stefansson Arctic Institute.

Arctic Energy Summit (2018a) *About*. Arctic Energy Summit 2017 website. Available at: www.arcticenergysummit.com/story/About (Accessed: February 17, 2019).

Arctic Energy Summit (2018b) *Past summits: about the Arctic Energy Summit*. Arctic Energy Summit 2017 website. Available at: http://arcticenergysummit.com/story/Past_Summits (Accessed: April 10, 2018).

AREA (2018) *Arctic renewable energy atlas* [Online]. Available at: http://arcticrenewable energy.org/home/about (Accessed: June 20, 2018).

ARR (2016) *Arctic Resilience Report* [Online]. Available at: http://hdl.handle.net/11374/1838 (Accessed: January 4, 2018).

Boström, M. (2012) "A missing pillar? Challenges in theorizing and practicing social sustainability," *Sustainability: Science, Practice and Policy*, 8(1), pp. 1–14.

Cherp, A. and Jewell, J. (2014) "The concept of energy security: beyond the four As," *Energy Policy*, 75, pp. 415–421.

Del Río, P. and Burguillo, M. (2008) "Assessing the impact of renewable energy deployment on local sustainability: towards a theoretical framework," *Renewable and Sustainable Energy Reviews*, 12, pp. 1325–1344.

Di Muzio, T. and Salah Ovadia, J. (2016) "Energy, capitalism and world order in IPE" in Di Muzio, T. and Salah Ovadia, J. (eds.) *Energy, capitalism and world order: toward a new agenda in international political economy*. Basingstoke: Palgrave MacMillan, pp. 1–19.

Emmerson, C. and Lahn, G. (2012) *Arctic opening: opportunity and risk in the High North*. Chatham House—Lloyd's Risk Insight Report [Online]. Available at: www.chathamhouse.org/publications/papers/view/182839 (Accessed: July 16, 2014).

Forbis, R. and Hayhoe, K. (2018) "Does Arctic governance hold the key to achieving climate policy targets?" *Environmental Research Letters*, 13(2).

Freudenburg, W. (1992) "Addictive economies: extractive industries and vulnerable localities in a changing world economy," *Rural Sociology*, 57(3), pp. 305–332.

Hannigan, J. (2015) *The geopolitics of deep oceans*. Cambridge: Polity Press.

Harsem, Ø., Eide, A. and Heen, K. (2011) "Factors influencing future oil and gas prospects in the Arctic," *Energy Policy*, 39, pp. 8037–8045.

IEA (2015) *Energy and climate change: world energy outlook special report* [Online]. Available at: www.iea.org/publications/freepublications/publication/WEO2015 SpecialReportonEnergyandClimateChange.pdf (Accessed: July 6, 2016).

IEA (2017) *World energy outlook 2017*: *executive summary*. Paris: International Energy Agency.

IEA (2018) *Global energy and CO2 status report 2017*. Paris: International Energy Agency.

Julien, H. (2008) "Content analysis" in Given, L. M. (ed.) *The SAGE encyclopedia of qualitative research*. Thousand Oaks: Sage, pp. 121–122.

Kristoferssen, B. (2014) *Drilling oil into Arctic minds? State security, industry consensus and local contestation*. Troms: Troms University Press.

Kruyt, B., Van Vuuren, D., de Vries, B.H.J.M. and Groenenberg, H. (2009) "Indicators for energy security," *Energy Policy*, 37(6), pp. 2166–2181.

Lempinen, H. (2017) *The elusive social: remapping the soci(et)al in the Arctic energy-scape*. Rovaniemi: Lapland University Press.

Lempinen, H. (2018) *Arctic energy and social sustainability*. London: Palgrave MacMillan.

Lempinen, H. and Cambou, D. (2018) "Societal perspectives on energy security in the Barents region" in Hossain, K. and Cambou, D. (eds.) *Societal security in the Arctic Barents region: environmental sustainability and human security*. London: Routledge, pp. 118–133.

Liuhto, K. (ed.) (2009) *The EU-Russia gas connection: pipes, politics and problems* [Online]. Available at: www.utu.fi/fi/yksikot/tse/yksikot/PEI/raportitjatietopaketit/Documents/Liuhto%200809%20web.pdf (Accessed: April 10, 2015).

Loe, J. and Kelman, I. (2016) "Arctic petroleum's community impacts: local perceptions from Hammerfest, Norway," *Energy Research & Social Science*, 16, pp. 25–34.

McGlade, C. and Ekins, P. (2015) "The geographical distribution of fossil fuels unused when limiting global warming to 2°C," *Nature*, 517, pp. 187–190.

Mitchell, J., Morita, K., Selley, N. and Stern, J. (2001) *The new economy of oil: impacts on businesses, geopolitics and society*. London: Earthscan.

Murphy, K. (2012) "The social pillar of sustainable development: a literature review and framework for policy analysis," *Sustainability: Science, Practice and Policy*, 8(1), pp. 15–29.

Owen, N., Inderwildi, O. and King, D. (2010) "The status of conventional world oil reserves: hype or cause for concern," *Energy Policy*, 38(8), pp. 4743–4749.

Paillard, C. (2010) "Russia and Europe's mutual energy dependence," *Journal of International Affairs*, 63(2), pp. 65–84.

Pickering, M. (2004) "Qualitative content analysis" in Lewis-Beck, M., Bryman, A. and Liao, T. (eds.) *The SAGE encyclopedia of social science research methods*. Thousand Oaks: Sage, p. 890.

Rasmussen, O.R. and Roto, J. (2011) *Megatrends*. Copenhagen: Nordic Council of Ministers.

Sidortsov, R. (2016) "A perfect moment during imperfect times: Arctic energy research in a low-carbon era," *Energy Research & Social Science*, 16, pp. 1–7.

Sidortsov, R. and Sovacool, B. (2015) "Left out in the cold: energy justice and Arctic energy research," *Journal of Environmental Studies and Sciences*, 5(3), pp. 302–307.

UNFCCC (2016) *Paris Agreement* [Online]. Available at: https://unfccc.int/files/essen tial_background/convention/application/pdf/english_paris_agreement.pdf (Accessed: May 14, 2017).

USGS (2009) *Assessment of undiscovered petroleum resources of the Barents Sea Shelf*. United States Geological Survey fact sheet [Online]. Available at: http://pubs.usgs.gov/fs/2009/3037/pdf/FS09-3037.pdf (Accessed: August 5, 2017).

Vallance, S., Perkins, H. and Dixon, J. (2011) "What is social sustainability? A clarification of concepts," *Geoforum*, 42, pp. 342–348.

Vanclay, F. (2002) "Conceptualising social impacts," *Environmental Impact Assessment Review*, 22, pp. 183–211.

Vanclay, F. (2003) "International principles for social impact assessment," *Impact Assessment and Project Appraisal*, 21(1), pp. 5–11.

WCED (1987) *Our common future*. United Nations World Commission on Environment and Development [Online]. Available at: www.un-documents.net/our-common-future.pdf (Accessed: January 14, 2012).

Empirical references: Arctic Energy Summit final reports

AES (2010) *The Arctic Energy Summit final report and technical proceedings: the Arctic as an emerging energy province* [Online]. Available at: https://oaarchive.arctic-council.org/handle/11374/36 (Accessed: October 10, 2018).

AES (2013) *The Arctic Energy Summit executive summary: richness, resilience, responsibility* [Online]. Available at: http://arcticenergysummit.com/files/executivesummary2013_web_final-20161016035906.pdf (Accessed: October 10, 2018).

AES (2015) *The Arctic Energy Summit executive summary: security and affordability for a resilient North* [Online]. Available at: http://arcticenergysummit.com/files/executivesummary2015_web_final-20161016035755.pdf (Accessed: October 10, 2018).

AES (2017) *2017 Arctic Energy Summit Finland final report: remote, renewable, responsible. Energy leadership in the Arctic* [Online]. Available at: www.sdwg.org/wp-content/uploads/2018/02/AES_Finland_final_report_proof2.pdf (Accessed: October 10, 2018).

6 Socially responsible investments (SRIs) in the European Arctic

New pathways for global investors to outperform conventional capital investments?

Adrian Braun

Introduction

Global challenges linked to ecosystem services, carbon footprints and the pollution of the natural environment are increasing for several reasons. A growing world population, a warming climate and international financial systems that have determined economic growth as "the maxim" all threaten the natural environment and thus jeopardize the living conditions of people in the Arctic and elsewhere on the globe (Porter and Kramer, 2006). These three causes are not entirely separate from each other but are rather intricately intertwined in the global economy (International Resource Panel report, 2011). Even though there are a myriad of theories and approaches on how to decouple economic growth from extended carbon footprints and negative impacts on the natural environment, in practice such decoupling is only marginally successful on a global scale (IPCC, 2018). Numerous industrial as well as developing countries have failed to achieve predetermined limits on greenhouse gas emissions to keep the projected negative impacts of climate change on an adaptable scale for humanity and the global ecosystems (Narayan et al., 2016; International Resource Panel report, 2011).

The early 2000s were characterized by a paradigm shift in terms of who should initiate action to preserve the natural environment. The duties of the industries have increased in this regard, largely as a result of higher pressure by stakeholder communities (Carroll and Buchholtz, 2003; Husted and Milton de Sousa-Filho, 2017). The concept of corporate social responsibility (CSR) has gained enormous acceptance in the global (business) community (Decker and Sale, 2009), creating powerful incentives for industrial companies to find eco-friendly solutions. Stakeholder conflict prevention, efficiency gains, cost reductions and finding market niches are just some of the reasons that motivate corporate decision-makers to implement CSR in business strategies (Carroll and Buchholtz, 2003; Klick, 2009).

In light of neoliberal thinking that markets can regulate themselves and political intervention is marginally needed or not needed at all, it is of interest how industries can preserve ecosystems and what activities they can enforce to achieve it

(Tennberg et al., 2014). Every business and every corporate project need the input of capital. The required capital can be achieved by reaching out to international capital markets (Luxembourg Bourse, no date). One way is the issuance of corporate shares on the counter (stock exchange) or off the counter (private trading). Diverse financial instruments are available to buy into a stock corporation, and diverse conditions, responsibilities and frameworks are linked to each instrument. Before explaining the novel connection of financial products and environmental and social impacts further, this chapter will first discuss the role of investors in global finance and, more specifically, the theme of the SRIs.

Today, the global community shares concerns of sustainable development and how corporate actors approach their social responsibilities. Scholars, activists and consumers pay increased attention to the role and performance of investors in this discourse (Porter and Kramer, 2006). Investors' performances and investment impacts on the ecological footprint and social and cultural sustainability are more transparent than in past decades. New media channels (such as blogs, vlogs, podcasts, discussion forums) and the increasing practices of nonfinancial business reporting are notable reasons for this development.

In this context, the term ESG was coined in 2005, abbreviating the three elements of environmental, social and governance (Principles for Responsible Investment, no date). ESG meets the idea of CSR in many ways with a distinct actor's point of view (Sanches Garcia et al., 2017). While CSR addresses industrial actors, the term ESG is linked to socially responsible investments (SRIs) and thus embraces the responsibilities of investors (Principles for Responsible Investment, no date). The concept of SRIs stretches the criteria of investment decisions, going beyond the unidirectional goal of creating profits. Institutional investors in particular, such as pension funds, banks and insurance groups, came into focus as they in many cases serve societal needs. It is thus of societal interest where the capital ends up. Pressure arises if investments move toward projects or enterprises that diminish the quality of life for the society and/or the natural environment. Tobacco, arms and gambling industries are examples of low ESG performance. Pressure is rising on their investors (Lee and Moscardi, 2017) while investments in philanthropic initiatives can improve stakeholders' lives and the relations between the "finance community" and the "interested parties," including local communities, indigenous groups, authorities, NGOs, customers and sub-investors (Husted and de Sousa-Filho, 2017; Carroll and Buchholtz, 2003).

It is similarly useful to evaluate the relevance of SRIs by following the example of regional developments and regional economies in the European Arctic. Global megatrends have a relevance in the most Northern latitudes in Europe. A globally warming climate plays a dominant role in this regard, with an average temperature rise in the Arctic twice the global average (IPCC, 2018; UNFCCC, no date). This wider attention on the European Arctic with its challenges and role in the sustainability discourse have put the area and its industries in a position that requires consideration of ESG characteristics (Kokko et al., 2014; Filkova and Frandon-Martinez, 2018).

SRIs can be effectively used to finance projects and initiatives to accomplish ecological or societal goals. Figure 6.1 depicts a few prominent examples of purposes that issuers of SRI products could have in mind that should be financed with the "fresh green capital." These purposes are very often associated with climate goals – not only in the Arctic context – to serve mitigation and adaptation strategies to counteract the negative impacts of a warming globe. The construction of energy-efficient buildings, factories and vehicles by simultaneously decreasing pollution levels and greenhouse gas emissions is a notable effort that is in line with SRI purposes and enables positive effects in the mid- and long-term perspectives.

This chapter considers the topic of green investments from two specific angles. One is the need and the potential in the European Arctic markets to invest socially responsibly. This means that territorial framework conditions and demographics may be relevant in raising capital: for example, to finance clean energy projects, low-carbon production facilities or electro-mobility in Arctic logistics. I will outline the six European Arctic markets and the potential capital requirements to strengthen their ecological and societal performances. The other focus is on the opportunities of the six European Arctic countries to raise capital on the international financial markets to follow "green" purposes. I will also discuss what kind

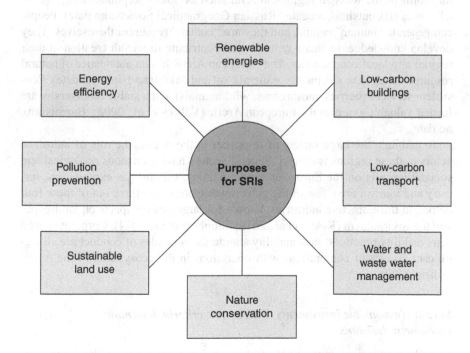

Figure 6.1 Purposes for raising capital in the framework of socially responsible investments (e.g. by issuing climate bonds or sustainability bonds or establishing ESG funds)

of green capital has already been allocated to the European Arctic, which actors have been dominant in utilizing them and which actors are still reluctant to apply modern financial instruments.

Background

This study links two discourses: socially responsible investments (green investments), which get connected to the European Arctic, and the requirements to strengthen a vulnerable Arctic ecosystem and to support social and cultural sustainability in the Northern latitudes. It is of interest how a rising demand of practices toward sustainable development affects the decisions of the investment community. The focus area is the European Arctic and Sub-Arctic, which embraces territories in Denmark (Greenland), Iceland, Norway, Sweden, Finland and Northwest Russia.

The European Arctic: a myriad of natural and "human" resources

The European Arctic is rich in resources and not only the natural kind (Barentsinfo, no date). The population figures in the most Northern latitudes in Europe are much higher than in other Arctic territories, such as in most of Canada's far North or the Russian regions in Asia such as Sakha Republic, Magadan or Chukotka (Barentsinfo, no date; Russian Geographical Society, no date). People can generate "human" capital, and they are "human" resources themselves. They develop knowledge, maintain cultures and contribute to wealth creation in their regions and local communities. The European Arctic has an abundance of natural resources, such as wood, metals, minerals, oil and gas (Barentsinfo, no date). Ecosystem services (berries, mushrooms, wild animals) and a stable biodiversity are further valuable assets in the European Arctic (Välkky et al., 2008; Barentsinfo, no date).

To maintain the large variety of resources in the Arctic, the role of industrial actors in these regions is crucial. Several sectors have enormous ecological and societal impacts on the European Arctic. Mining, oil and gas exploitation, forestry and tourism are a few of the most notable ones, and three out of these four belong to the extractive industries, known for large-scale impacts on landscapes and the environment (Kokko et al., 2014; Tennberg et al., 2014). Corporate social responsibility practices, sustainability standards and codes of conduct are all significant aspects of corporations with operations in the ecosystems of the Arctic (Klick, 2009).

Socially responsible investments (SRIs): new criteria determine investment decisions

Socially responsible investments follow multiple purposes at the same time (World Economic Forum, 2015). Just as with any conventional investment, the investors are seeking a return on their investment (profits), capital security and/or dividends. However, SRIs target the achievement of a certain ecological or social benefit (Choi, 2017; Barclays, 2018; Kidney and Boulle, 2015). Those investment projects can strive to reduce greenhouse gas emissions in production processes, to improve waste management or to prevent industrial pollution. Examples of societal benefits include health and safety policies to protect the workforce against accidents and support for local communities (e.g. education, pensions). SRIs are largely intended to strive for sustainable development wherever the capital is allocated. In this context, the ESG approach was coined in 2005, covering three elements: environmental, social and governance (Thomson Reuters, 2017; Gitman et al., 2009; Lee and Moscardi, 2017; Barclays, 2018).

Investors may turn their backs on conventional investments and focus more strongly on SRIs for multiple reasons. Investment criteria can embrace both negative and positive aspects of the investment target. The negative criteria might include the investor's disapproval of specific production methods (high pollution levels, high resource consumption, child labor, unfair wages) or disapproval of the product or service itself (e.g. weapons, nuclear power, narcotics). With the help of ESG analyses, a "socially responsible" investor can either invest or divest capital on the basis of corporate ESG performance (Barclays, 2018; Gitman et al., 2009). ESG analyses can in equal measure highlight positive results toward the three ESG dimensions. Investments into cleaner production or improved working and living conditions are good examples in this context (Sanches Garcia et al., 2017; Lee and Moscardi, 2017).

Investors do not usually provide capital carelessly. One or more benefits are expected from the investment, including financial profit, and socially responsible investments are no exception. Investors will not buy financial products to serve cultural, social and ecological benefits if it means that they themselves face certain capital losses. Thus, in order to be successful and gain acceptance on the capital markets, "green" investments must be profitable (at least in the long run) and allow effective enforcement of sustainable development at the same time (Thomson Reuters, 2017; Choi, 2017). Analysts of financial institutions as well as economists and social scientists have examined whether ESG investments level up with conventional investments or if they are even able to outperform them (Sanches Garcia et al., 2017). The findings are contested. While some traders and analysts argue that green investments often entail a margin net loss in a portfolio, others claim that responsible investments often outperform conventional investments (Khan et al., 2016; Choi, 2017). More consensus has been achieved in terms of the stability of ESG investments. They seem to be more resilient toward "big loss" threats on the international markets (Sanches Garcia et al., 2017; Khan et al., 2016). Fluctuations and volatility entail a risk, and the trading community seeks to avoid risks. An investment with strong ESG performance is less risky on big market capitalization losses because these investments are rarely linked to ecological scandals, court cases, strikes, societal conflicts, negative media campaigns and other threats that cause price drops on the stock exchanges and trading floors (Barclays Investment Bank, 2018; Lee and Moscardi, 2017).

SRIs and financial products: the rise of climate bonds

In 2016, China initiated a program to establish a Green Financial System; several corporate and government actors in China drew up guidelines for this framework (Meng et al., 2018). The financial instruments described in the outcomes of this work comprise what is nowadays available on the global markets in terms of financial products to commit to SRIs. In addition to climate bonds, these are green stock indices (e.g. derivatives), green credits, green development funds, green insurance and carbon finance (Nassiry, 2018).

The European Arctic is linked to diverse developments and challenges. Out-migration of the youth and urbanization are two developments that threaten regional societies. The dominant challenge that has caught attention over the past two to three decades on the entire globe is climate change (IPCC, 2018; UNF-CCC, no date). The rise of global sea levels due to melting polar ice tops the list of global warming threats. The UNFCCC and its annual Conferences of the Parties (COP) with the resulting Kyoto Protocol in 1997 (COP 3) and the Paris Agreement in 2015 (COP 21) have contributed to a wider awareness of the issue.

It is therefore not surprising that the most elaborate and the most accepted financial product so far suitable for SRIs links to the topic of climate change (World Economic Forum, 2015). Climate bonds (green bonds) issued on the international capital markets totaled 135 billion euro in 2017 and 167 billion euro in 2018 and are estimated to amount to 250 billion euro by the end of 2019 (Climate Bonds Initiative, 2018). Mark Carney, governor of the Bank of England, described an essential problem in the ways in which the finance sector can address climate change issues. In his speech "Breaking the tragedy of the horizons: climate change and financial stability" (2015), Carney underlines the dilemma that the majority of investments are short term, while it takes long-term initiatives over years, even decades, to counteract the negative impacts of climate change. How can investors who offer their capital to a specific financial instrument (e.g. ordinary or preferred company shares) short term (for a few minutes, hours or days) before withdrawing (selling) it again follow a long-term purpose such as mitigating or adapting to global warming? With this investment strategy, it is very difficult, because the incentive is missing. Day trading (establishing and clearing all investment positions within one trading day) is done at such a pace that the issue that matters most is the stock price development. Product features, company reputation and ESG factors do not play a role in these short-term investments as the investors do not have time to analyze these characteristics. Seen from this viewpoint, day trading possesses a rather unsustainable character.

The raising or investing of capital and the utilization of financial products on the international capital markets are for many actors associated with worries of committing significant mistakes that might cause substantial losses. These concerns are largely connected to a lack of knowledge about the procedures on trading floors and stock exchanges and the myriad of financial products available, each with its own characteristics and legal requirements. It is necessary to understand that even experienced investment portfolio managers cover only one or a few specific segments and strategies of trading. There are specialists in "swaps," "futures," "derivatives," bonds and other instruments, and the specialization within these segments is further narrowed down to regional markets, sectors, currencies and other features.

So, to focus on climate bonds, one needs to know and learn about this specific product. This applies to bond buyers in particular, whereas bond issuers can work with an investment bank or financial advisory firm and do not have to know the specific details of the bond market in depth. They just have to know the required capital and need a plan on how this capital is going to be used to follow the defined "climate purpose."

A practical example of how climate bond issuance works for an actor affiliated with the Arctic is outlined in the following. We assume an Arctic city with around 50,000 inhabitants and a stable local economy and society, functioning local industries, normal debt levels and no significant population decline. The city center renewal project plan, which has been ready for quite some time, foresees the renovation of several buildings and the construction of a new town hall. All the efforts shall be in line with the best available solutions in terms of low consumption of water and energy and low greenhouse gas emissions. The greater idea behind this is fighting against climate change and contributing to environmental programs. The realization of the project depends on successful financing. Issuing a climate bond is the identified solution to access the required funding. Simplified, what needs to be clearly defined is the purpose, the required capital (amount issued), the project goals, the maturity (the time span of the bond), the coupon (the interest rate for bondholders' investments), the jurisdiction and market place and the long-term financial plan to cover the bond. Going back to the example of the Arctic city center renewal, this could require 80 million euro with a maturity of over 20 years and a coupon of 2.5 percent per annum. Once the climate bond is issued (for example, on the Luxembourg Stock Exchange [Luxembourg Bourse, no date]), large-scale investors (banks, pension funds, insurance groups, countries) and small- and medium-size investors can buy bonds until the overall required capital is raised. Bondholders are now eligible to receive interest and can trade their bonds on the marketplaces. The climate bond is a debt security (just like any conventional bond). In our example, the Arctic city is the debtor, and the bondholders are the creditors in these financial interrelations.

What, then, is the difference between climate bonds and other financial instruments? In terms of their basic technical features, they are no different from any other bond (Clapp et al., 2016). Bonds are debt securities and are issued by an actor (in most cases, large-scale companies or governments) to raise capital that fulfills a certain purpose. By buying the bond, the holders make a loan to the issuer. The investor can use the bond for speculation or they can wait up to the maturity date (closing date of the bond) to get their investment back, meanwhile receiving interest. Bondholders have an advantage over stockholders; in case of bankruptcy, they are considered as lenders and are paid first from the remaining insolvency assets. The disadvantage is that bondholders do not have voting rights or other forms of co-determination. Moreover, climate bonds tend to be issued with flat pricing: no complicated regulations and options go along with the purchase of a bond (Lee and Moscardi, 2017; Filkova and Frandon-Martinez, 2018). The issuers of climate bonds face the disadvantage of higher transaction costs than conventional bonds, due to the requirements of monitoring and reporting on the use of proceeds (Clapp et al., 2016). Image benefits, green marketing opportunities and attraction of investors who especially focus on ESG performances of financial products may outweigh these disadvantages for the issuer. For example, China and its governmental and corporate decision-makers have discovered climate bonds as a financial instrument to be a pillar in their continuous pursuit of rapid economic growth. Since 2017, China has issued climate bonds exceeding 30 billion euro (Meng et al., 2018; Climate Bonds Initiative, 2018).

Climate bonds in the European Arctic: toward sustainable development

The majority of Northern European states have already issued climate bonds. Sweden is the most active country in this regard, having issued bonds for more than 10 billion euro as of 2018, followed by Norway (issues totaling 2.7 billion euro), Denmark (Greenland) (2.3 billion euro) and Finland (around 1 billion euro). Iceland has not yet issued a climate bond. There are a couple of climate bonds denominated in Russian rubles, but none of them belong to a Russian corporation, governmental actor or other Russian affiliation (World Bank, 2017; Climate Bonds Initiative, 2018). The few European markets and their status on climate bonds is further assessed in the following.

Iceland

Iceland has not yet issued a climate bond, but its geographic location (the Northern shore of the mainland is a few dozen kilometers south of the Arctic Circle) and the threat of climate change translate into incentives to issue climate bonds in the future. Iceland is active in the fisheries and tourism sectors, which both feature prominently in the sustainable development discourse. The Climate Bonds Initiative has found that issuances of corporate bonds are popular in Iceland and that housing and real estate may have a strong potential for issuing climate bonds in the green housing sector (Filkova and Frandon-Martinez, 2018).

Norway

In January 2018, the Norwegian SpareBank alliance issued a 1 billion euro climate bond with a duration of seven years (maturity date in 2025). The purpose of this bond is the development of low-carbon buildings in Norway to contribute to the country's overall pursuit of the mitigation of greenhouse gas emissions and the negative impacts of climate change. The strategy is to allocate the net proceeds of SpareBank's climate bond to a loan portfolio linked to diverse mortgages to allow the construction of energy-efficient and low-carbon buildings. SpareBank has established criteria (with an advisory firm) that construction companies have to fulfill in order to be eligible to access the raised ("green") capital. The key criterion is the energy-efficiency level of the planned buildings. The SpareBank climate bond is one of the few to have received certification by the Climate Bonds Initiative (certified climate bonds), with approval of an external verifier (usually a consulting or sustainability advisory firm). While Sweden has issued the largest amount of climate bond capital among the countries in the European Arctic, Norway was the first Northern European country to issue a climate bond in 2018. Already in May 2010, KBN Kommunalbanken had issued two bonds worth 85 million euro to finance initiatives and projects that contribute to a low-carbon society (Kommunalbanken Green Bonds, 2018; Filkova and Frandon-Martinez, 2018).

Sweden

Sweden's first issuance of a climate bond took place in 2013, when the city of Gothenburg (2015) became the first city issuer on the globe. They raised capital in the amount of 57 million euro. This was the leadoff to a process of local governments' reaching out to the capital markets to achieve long-term finance in the framework of SRIs (Filkova and Frandon-Martinez, 2018). In the past few years, the city of Gothenburg has issued three more climate bonds. All the city's climate bonds raised capital to finance diverse climate-related projects to advance climate-resilient growth and a low-carbon economy (UNFCCC). The incentive for Gothenburg to seek capital was based on their own ecological problems, which were at a peak in the late 1980s (City of Gothenburg, 2015). Precise environmental goals and a climate strategy helped show a path out of the misery, and climate bonds were an applicable financial tool on the way to the solution. The raised (socially responsible) capital was used to establish a fleet of electric cars in the urban environment, a biogas energy project and a heating grid that allows specific vessels to dock at the Gothenburg port and run the systems without burning carbon-intensive fuels (City of Gothenburg, 2015). Many other Swedish local governments (cities and municipalities) have followed the example of Gothenburg, including Malmö, Västerås, Örebro and Östersund (Climate Bonds Initiative, 2018; Nassiry, 2018). From the perspective of corporate issuers, the Vasakronan real estate company has been very active in issuing numerous bonds raising capital of more than 585 million euro, largely targeted to building modern houses with low energy consumption (Nassiry, 2018).

Denmark (Greenland)

Ørsted A/S is the largest energy supplier in Denmark and a major investor in multiple offshore wind energy projects in Europe and Asia. The decision-makers of this power supplier have identified climate bonds as the appropriate tool to finance renewable energy projects (Ørsted A/S Company Announcement, 2017; Wienberg, 2017). As of 2018, Ørsted holds a market share of 16 percent of offshore wind energy and already has more than 1,000 wind turbines installed in the framework of their projects. To finance at least parts of these projects, the amounts issued by Ørsted A/S exceeded 1.25 billion euro as of 2018 (Wienberg, 2017). Unlike Sweden, the Danish local governments (cities/municipalities) have hardly issued any climate bonds to this date. The vast majority available on the market are corporate bonds. Considering that Greenland belongs to Denmark and is, due to its geographical location in the Arctic, particularly vulnerable to impacts of climate change that may threaten the entire planet, it is relevant to encourage actors to issue climate bonds that follow specific climate goals in Greenland. This is all the more important given that the mining sector has several exploration projects underway in Greenland, and large-scale exploitation of metals and minerals at several locations is likely in the future (Geological Survey of Denmark and Greenland: Mineral Resources, no date).

Finland

Finland is involved in industries that have high carbon outputs into the atmosphere, but the first climate bond in Finland was issued comparatively late, in October 2016. Municipality Finance (MuniFin), a Finnish credit institution, started its green finance program in 2016 (MuniFin Municipality Finance, 2018), reaching out to the international capital markets to provide loans for projects in Finland, which have to fit in one or more of the following categories: renewable energies, energy efficiency, sustainable public transport, waste management, water management, sustainable buildings and nature conservation/environmental management (MuniFin Municipality Finance, 2018; Nassiry, 2018). Altogether, MuniFin has issued around 900 million euro to enable financing of projects in these categories (maturity of five years). The MuniFin strategy also leads to the fact that the allocation of climate bonds capital toward climate change adaptation is relatively higher in Finland than in the other European Arctic countries. Adaptation proceeds of use in Finland are up to 17 percent of the investment share, while the runner-up in this category, Sweden, allocates only around 3 percent to adaptation purposes. The overall amount of climate bonds issued in Finland is low. As of 2018, only three other climate bonds (in addition to those launched by MuniFin) have been issued, two by local government funding agencies and one by Fingrid, to enlarge the share of renewable energies in the Finnish energy mix (Fingrid Green Financing, no date; Filkova and Frandon-Martinez, 2018).

Russia

The multinational Russian aluminum producer RUSAL is preparing (including necessary "green" certification) to issue a climate bond – without confirming that it is going to happen. Financial media sources reported in January 2018 that RUSAL is going to issue a climate bond worth more than 400 million euro (Hale, 2018). Metal production requires considerable inputs of natural resources and energy and is linked to numerous ecological challenges. These include climate-relevant issues such as greenhouse gas emissions from smelting, production and transportation (Kokko et al., 2014). Low-carbon technologies and other sustainable practices could be financed for RUSAL by issuing a climate bond to improve the overall environmental performance of the company. As the Russian economy is largely dependent on extractive industries (oil, gas, mining, forestry), the same opportunities are applicable to other companies operating in these sectors. The World Bank has issued diverse Russian ruble climate bonds to support climate-related projects in developing countries (without nomination of specific countries). These bonds are listed in Luxembourg and have maturities of around ten years (Luxembourg Bourse). Russian actors themselves can try to get access to this capital by implementing climate projects in Russia. In Northwest Russia (e.g. in the Murmansk region, Arkhangelsk, the Komi Republic and the Republic of Karelia), there are numerous infrastructural projects underway on- and offshore (e.g. railway construction, icebreaker fleet realization) that have the potential to receive green investments and consequently improve the sustainability performances of the regional assets (Staalesen, 2016).

Conclusion

Socially responsible investments (SRIs) are on the rise in the European Arctic as well as on the global markets. Over the past decades, the global finance sector has developed a wealth of financial instruments to raise and/or invest capital. SRIs are a novel platform to stir the creativity of business actors and the investment community. Numerous bonds, funds and other securities are already available for social responsibility purposes. One of the most accepted and effective "green" financial instruments is climate bonds. They entered the landscape of Northern Europe in 2010, and a continuous number of issuers have shown up every year since Kommunalbanken's (Norway) kickoff.

Sweden and Norway both have actors from industries and governments that are active in issuing climate bonds. These two countries, among the European Arctic states, have raised the highest capital amounts by issuing climate bonds. Municipalities in Finland and Denmark (Greenland) are yet to perceive the opportunities to reach out to capital markets to strive for projects related to ecological and social concerns. Finnish Lapland and Greenland (Denmark) have recently received plenty of industrial investments in, for instance, minerals and metals exploration. Local governments in these areas have climate ideas and proposals that are hard to implement due to a lack of financial resources. "Green" investments and the issuance of "green" financial products can be a way of realizing these ideas. Gothenburg, Malmö and Östersund in Sweden have shown that it is possible, and Gothenburg issued several more bonds after its initial success. In Iceland and Russia, the SRI field is undeveloped both for governmental and corporate actors. However, the potential is there, as in every other European Arctic country, given the major industries and the geographic location of these states.

China can become a crucial actor in this discourse as well. The country has issued a huge number of climate bonds and has shown great interest in the Arctic over the past decade. A Chinese climate bond that covers a specific purpose in the European Arctic to counteract or adapt to global warming has still to come.

Overall, the discourses of sustainable development and corporate social responsibility that have largely determined the business and political agendas of the past two decades have spread out to the investment community. As the impacts of climate change and the challenges of population growth and urbanization are unlikely to slow down in the near future, the relevance of SRIs as novel instruments to finance counteraction against unwanted impacts will grow.

References

Barclays Investment Bank (2018) *Sustainable investing and bond returns* [Online]. Available at: www.investmentbank.barclays.com/content/dam/barclaysmicrosites/ibpublic/documents/our-insights/esg/ ontente-sustainable-investing-and-bond-returns-3.6mb.pdf (Accessed: June 30, 2018).

Barentsinfo (no date) *Facts on the regions* [Online]. Available at: www.barentsinfo.org/Contents/Statistics/Population (Accessed: July 3, 2018).

Carney, M. (2015) *Breaking the tragedy of the horizon: climate change and financial stability*. Speech given at Lloyd's of London [Online]. Available at: www.bankofengland.co.uk/-/media/boe/files/speech/2015/breaking-the-tragedy-of-the-horizon-climate-change-and-financial-stability.pdf?la=en&hash=7C67E785651862457D99511147C7424FF5EA0C1A (Accessed: July 2, 2018).

Carroll, A.B. and Buchholtz, A.K. (2003) *Business: ethics and stakeholder management*, 7th edn. Cincinnati: South Western Pub.

Choi, A. (2017) *Why sustainable companies can outperform* [Online]. Available at: www.morganstanley.com/access/why-sustainable-companies-can-outperform (Accessed: July 3, 2018).

City of Gothenburg (2015) *Environmental program* [Online]. Available at: http://international.goteborg.se/green-gothenburg (Accessed: November 2, 2018).

Clapp, C., Alfsen, K.H., Francke Lund, H. and Pillay, K. (2016) *Green bonds and environmental integrity: insight from CICERO second opinions*. CICERO Policy Note. Oslo: CICERO Center for International Climate and Environmental Research.

ClimateBondsInitiative(2018)*Certifiedclimatebonds*[Online].Availableat:www.climatebonds.net/standard (Accessed: July 10, 2018).

Decker, S. and Sale, C. (2009) "An analysis of corporate social responsibility, trust and reputation in the banking profession" in Idowu, S.O. and Filho, W.L. (eds.) *Professionals' perspectives of corporate social responsibility*. Berlin and Heidelberg: Springer-Verlag, pp. 135–156.

Filkova, M. and Frandon-Martinez, C. (2018) *The green bond market in the Nordics* 2018 [Online]. Available at: www.climatebonds.net/resources/reports/green-bond-market-nordics (Accessed: July 3, 2018).

Fingrid (no date) *Green financing* [Online]. Available at: www.fingrid.fi/en/pages/investors/financing/green-financing/ (Accessed: July 10, 2018).

Geological Survey of Denmark and Greenland (no date) *Mineral resources* [Online]. Available at: www.eng.geus.dk/mineral-resources/ (Accessed: July 10, 2018).

Gitman, L., Chorn, B. and Fargo, B. (2009) *ESG in the mainstream: the role for companies and investors in environmental, social, and governance integration*. Business for Social Responsibility. New York: BSR Network.

Hale, T. (2018) "The green bond that wasn't," *Financial Times Alphaville* [Online]. Available at: https://ftalphaville.ft.com/2018/01/24/2198049/the-green-bond-that-wasnt (Accessed: July 10, 2018).

Husted, B.W. and Milton de Sousa-Filho, J. (2017) "The impact of sustainability governance, country stakeholder orientation, and country risk on environmental, social, and governance performance," *Journal of Cleaner Production*, 155, pp. 93–102.

Intergovernmental Panel on Climate Change (IPCC) (2018) *Global warming of 1.5 °C*. IPCC Special Report [Online]. Available at: www.ipcc.ch/report/sr15/ (Accessed: November 10, 2018).

International Resource Panel (2011) *Decoupling natural resource use and environmental impacts of economic growth*. United Nations Environment Programme. A Report of the Working Group on Decoupling to the International Resource Panel [Online]. Available at: https://www.resourcepanel.org/reports/decoupling-natural-resource-use-and-environmental-impacts-economic-growth (Accessed: September 17, 2019).

Khan, M., Serafeim, G. and Yoon, A. (2016) "Corporate sustainability: first evidence on materiality," *The Accounting Review*, 91(6), pp. 1697–1724.

Kidney, S. and Boulle, B. (2015) "The opportunity for bonds to address the climate finance Challenge" in Wendt, K. (ed.) *Responsible investment banking: risk management frameworks, sustainable financial innovation and softlaw standards*. Cham: Springer, pp. 575–599.

Klick, M. (2009) *The political economy of corporate social responsibility and community development: a case study of Norway's Snøhvit natural gas complex*. Lysaker: Fridtjof Nansen Institute.

Kokko, K., Oksanen, A., Hast, S., Heikkinen, H.I., Hentilä, H., Jokinen, M., Komu, T., Kunnari, M., Lépy, É., Soudunsaari, L., Suikkanen, A. and Suopajärvi, L. (2014) *Sound mining in the north: a guide to environmental regulation and best practices supporting social sustainability*. DILACOMI Project (Different Land-Uses and Local Communities in Mining Projects).

Kommunalbanken(KBN)(2018)*Greenbonds*[Online].Availableat:www.kommunalbanken. no/en/funding/kbn-green-bonds (Accessed: July 5, 2018).

Lee, L-E. and Moscardi, M. (2017) *MSCI-Morgan Stanley capital international 2017: ESG trends to watch*. MSCI Report.

Luxembourg Bourse (no date) *The sustainable finance platform* [Online]. Available at: www.bourse.lu/green (Accessed: July 5, 2018).

Meng, A.X., Lau, I. and Boulle, B. (2018) *China green bond market 2017* [Online]. Climate Bonds Initiative and the China Central Depository and Clearing Co. Ltd (CCDC). Available at: www.climatebonds.net/resources/reports/china-green-bond-market-2017 (Accessed: July 5, 2018).

Municipality Finance (MuniFin) (2018) *Green finance exceeds EUR 1.0 billion* [Online]. Available at: www.munifin.fi/recents/news/2018/01/02/green-finance-exceeds-eur-10-billion (Accessed: July 7, 2018).

Narayan, P.K., Saboori, B. and Soleymani, A. (2016) "Economic growth and carbon emissions," *Economic Modelling*, 53, pp. 388–397.

Nassiry, D. (2018) *Green bond experience in the Nordic countries* [Online]. ADBI Working Paper 816. Tokyo: Asian Development Bank Institute. Available at: www.adb.org/publications/green-bond-experience-nordic-countries (Accessed: July 6, 2018).

Ørsted A/S Company Announcement (2017) *Ørsted issues green bonds* [Online]. Available at: https://orsted.com/en/Company-Announcement-List/2017/11/1645531 (Accessed: July 5, 2018).

Porter, M.E. and Kramer, M.R. (2006) "Strategy and society: the link between competitive advantage and corporate social responsibility," *Harvard Business Review*, December, pp. 78–94.

Principles for Responsible Investment (no date) *ESG issues* [Online]. Available at: www.unpri.org/esg-issues (Accessed: July 5, 2018).

Russian Geographical Society (no date) *Population statistics in the Arctic* [Online]. Available at: https://arctic.ru/population (Accessed: July 3, 2018).

Sanches Garcia, A., Mendes-Da-Silva, W. and Orsato, R.J. (2017) "Sensitive industries produce a better ESG performance: evidence from emerging markets," *Journal of Cleaner Production*, 150, pp. 135–147.

Staalesen, A. (2016) *These are Russia's top Arctic investments* [Online]. The Barents Observer. Available at: https://thebarentsobserver.com/ru/node/612 (Accessed: July 5, 2018).

Tennberg, M., Vola, J., Espiritu, A.A., Schwenke Fors, B., Ejdemo, T., Riabova, L., Korchak, E., Tonkova, E. and Nosova, T. (2014). "Neoliberal governance, sustainable development and local communities in the Barents Region," *Barents Studies*, 1(1), pp. 41–72.

Thomson Reuters ESG Scores (2017) [Online]. Available at: financial.thomsonreuters.com/esg (Accessed: July 7, 2018).

United Nations Framework Convention on Climate Change (UNFCCC) (no date) [Online]. Availableat:https://unfccc.int/climate-action/momentum-for-change/financing-for-climate-friendly/gothenburg-green-bonds (Accessed: July 5, 2018).

Välkky, E., Nousiainen, H. and Karjalainen, T. (2008) *Facts and figures of the Barents forest sector*. Helsinki: Finnish Forest Research Institute.

Wienberg, C. (2017) *The bond that really asks you to bet on the future of the planet* [Online]. Bloomberg. Available at: www.bloomberg.com/news/articles/2017-11-16/orsted-will-sell-first-green-bonds-as-part-of-move-away-from-oil (Accessed: July 5, 2018).

World Bank (2017) *World Bank launches Russian ruble green bonds* [Online]. Available at: www.worldbank.org/en/news/press-release/2013/06/13/world-bank-launches-russian-ruble-green-bonds (Accessed: July 3, 2018).

World Economic Forum (2015) *Arctic investment protocol* [Online]. Available at: http://www3.weforum.org/docs/WEF_Arctic_Investment_Protocol.pdf (Accessed: July 5, 2018).

7 Resources on the Arctic border
Views of the Finnish municipalities and the EU's cross-border program

Paula Tulppo

Introduction

What is development? Sotarauta defined development as "'good' public interest, which is intended to be improved by different kinds of developmental activities" (Sotarauta, 2015, p. 216). However, interpretations of "good interest" vary. The notions of "development" are socially determined by particular social groups and/ or interests in specific places and time periods (Pike et al., 2006, pp. 24–25).

In managing regional development, one or more actors participate in the development of a certain region, and traditionally decisions related to regional development have been made in national centers. In modern societies, regional management is normally not just one actor's responsibility. The aim is to direct an area's own resources and the actions of several organizations and external resources in the same direction and toward the right issues (Sotarauta, 2015, p. 215). This has to be done in cooperation with actors and organizations at different levels, and local views need to be heard.

More than ever, managing regional development now emphasizes sustainable development where environmental, economic, cultural and social aspects are integral elements to be considered (Katajamäki, 2011). Development is directed especially toward strengthening the regions' endogenous potential to develop, helping survival in the global economy and adjusting or exploiting exogenous powers to change. It is not always easy to identify the right resources to help a region develop. Furthermore, different actors might well have different opinions on this. (Sotarauta, 2015, pp. 215, 217.)

In general, discussions about Arctic regions and their resources for development often deal with natural resources such as forestry, ore and hydropower. However, if we consider the real and genuine development of these regions at the local level and the resources for development, natural resources are not enough to secure diverse and long-term development. The economic resources need a wider basis: ecological, social and cultural resources are equally vital. Without diverse and sustainable endogenous resources, long-term regional development is difficult.

This chapter analyzes resources for regional development on two different levels, from the perspective of six Finnish municipalities in the Arctic border region and from the vantage point of the EU Interreg V A Nord program, where the Arctic border area is only a part of a larger program region in northern Finland, Sweden and Norway. Both levels aim at supporting regional development: the municipalities on the local level and the EU's cross-border program in a wider area, which includes both growing cities such as Oulu and sparsely populated, remote areas with negative migration such as most of the municipalities in the Tornio Valley region. The needs and resources for development within the program region are multiple.

I will examine what kinds of resources for regional development have been recognized by the two levels, how diverse these resources are and how the discourses about local resources meet and differ.

Resources are understood as strengths and other items in promoting the development of a region. The research data includes the description and SWOT (Strengths, Weaknesses, Opportunities, Threats) analysis of the Interreg Nord program region (Interreg Nord, 2014), which is an appendix to the Joint Operational Program description of the Interreg V A Nord (2014), and the municipal strategies of the six Finnish municipalities of Tornio (2018), Ylitornio (2018), Pello (2015), Kolari (2017), Muonio (2018) and Enontekiö (2017).

The research region is located on the Finnish side of the Finnish-Swedish Arctic border (Map 7.1). Until 1809, there was no national border, and the region shared the same language, culture, religion and administration across the Tornio River and the Muonio River. The Hamina peace treaty between Sweden and Russia set the border along the Tornio and Muonio Rivers, and Finland became part of Russia. The united region was thus split, leaving relatives and friends on both sides of the border. Finland became independent in 1917, but the border between Finland and Sweden has been where it is today since 1809, and there has always been lively interaction across the border (Ruotsala, 2011, pp. 199–201).

The research area of the six municipalities of Tornio, Ylitornio, Pello, Kolari, Muonio and Enontekiö is sparsely populated, with a total of 37,635 inhabitants (in 2017). Tornio is the only town in the research area, with 21,928 inhabitants; all the other municipalities have a population of less than 5,000 each (Statistics Finland, 2018b).

Border regions are often far from national core centers, which causes lower synergy effects leading to lower development (Jańczak, 2018, p. 396). This border region also has its developmental concerns, such as decreasing population. Between 2007 and 2017, the population in the research area decreased by 2,021 inhabitants (Statistics Finland, 2018b). The number decreased in every municipality except in Kolari, where it increased slightly (Statistics Finland, 2018b).

Another worry is an aging population, which gives cause for concern in the whole of Finland, but even more so in the research area. In every municipality of the research area, the share of people in the population older than 65 years was bigger than in Finland in general, where the proportion amounted to 21.4 percent of the population in 2017. Of the six municipalities under study, the share of people aged 65+ was highest in Pello (37.3 percent) and in Ylitornio (35 percent). In the other four municipalities, the share was well under 27 percent (Statistics Finland, 2018b).

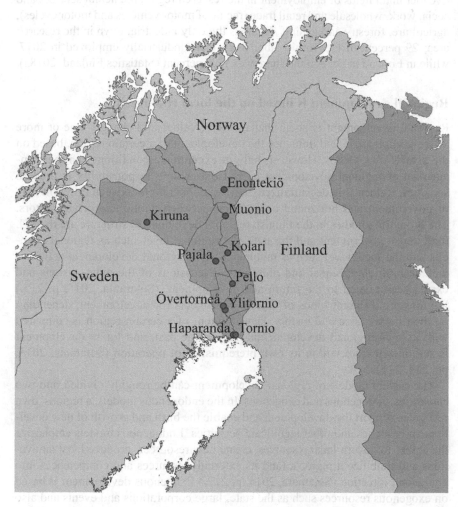

Map 7.1 The Arctic border region between Finland and Sweden

The unemployment rates in the research area are mainly above the Finnish average, but there are significant seasonal differences in some of the municipalities. The December 2017 unemployment rates in Muonio and Kolari were only 7.9 percent and 13 percent, respectively, but by June 2018 the rates had climbed to 19.9 percent in Muonio and 17 percent in Kolari. The overall Finnish unemployment rate at the time was about 11 percent. Such seasonal differences in Muonio and Kolari can most likely be explained by tourism, which is busier in winter than in summer. In 2018, 19 percent of the Muonio labor force and 12.7 percent in Kolari worked in the accommodation and catering services (Statistics Finland, 2018a).

Other main fields of employment in the research region are health services and social work, wholesale and retail trade (repair of motor vehicles and motor cycles), agriculture, forestry and fishing. Tornio is the only industrial town in the research area; 29 percent of the Tornio labor force were industrially employed in 2017, while in Finland in general the share was 12.6 percent (Statistics Finland, 2018a).

Regional development is based on the local resources

Regional development is about management, values and change. One or more actors participate in and influence the development of a certain region, based on the actor's own values. However, only in exceptional conditions can the management of regional development be just one actor's responsibility in modern societies. Rather, it is decentralized, vertical cooperation between different tiers of government and horizontal cooperation between public and private actors. The promoting bodies in the Finnish regional development structure are the EU, the Finnish government and the state, the regional level such as regional councils and the local level such as municipalities. Regional development activities include multidimensional and pluralistic discussions of the aims, visions and experiences related to the future and development. (Sotarauta, 2015, p. 215.) Regions are different kinds of objects or subjects for development, depending on their nature, size and history. Development of a certain region is compared with its own regional developmental path in the past, and future development is scrutinized from within its own prerequisites of operation (Sotarauta, 2015, pp. 215, 217).

The current models of regional development can be roughly divided into two categories, endogenous and exogenous. In the endogenous models, a regions' own resources support the development and enable the birth and growth of new development branches and other significant activities. Endogenous models emphasize the ability to govern local resources, create new resources, produce local innovations and mobilize a region's (and its external) resources and competences into one united direction (Sotarauta, 2015, p. 217.) Exogenous development is based on exogenous resources such as the state, large corporations and events and also on mobile capital, technologies, talent and knowledge (Sotarauta, 2015, p. 217; Tödtling, 2011, p. 340.)

According to Tödtling, a core idea in endogenous regional development is the understanding of development as a bottom-up process. Local and regional actors and initiatives play key roles; decision-making, policy competences and institutions are in order at both local and regional levels (Tödtling, 2011, p. 340; see also Jakola, 2016, p. 1815).

During the 21st century, the Finnish regional development policy has sought to support the region's own endogenous potential to develop. This does not mean neglecting the exogenous forces, but the basic idea is that development draws on the region itself. The strengthening of the region's own capacity to develop is a prerequisite to surviving in a global economy and adjusting to or exploiting external reform forces (Sotarauta, 2015, p. 217).

Endogenous local and regional development has responded to the problems and limits of a one-sided top-down or external development approach. It has brought attention to indigenous and endogenous forces and factors, including social and political processes, and it has emphasized the key role of local initiatives, entrepreneurship and innovation. However, there are also weaknesses and limitations in the endogenous development concept. Initially, it has focused too strongly on endogenous factors and actors, neglecting the fact that successful regional development is usually the result of both exogenous and endogenous forces. The concept of endogenous regional development has more or less tried to promote "islands of development" in a world of increasing social and economic interdependencies at all spatial levels (Tödtling, 2011, p. 340).

According to Pike et al. (2006), the focus of regional development has shifted from the quantity of development to issues of quality. This entails an emphasis on the impact of economic development on the natural environment and the constraints this places on development, but also stressed are questions of the quality of life (Pike et al., 2006, pp. 3–4). While the leading idea in regional development used to be that positive economic development would also engender other good aspects of life, it gradually became obvious that if the positive economic developments are achieved by exploiting environmental, social, cultural or other factors, the overall development can be negative (Pike et al., 2006). Today, regional development activities require that environmental, economic, cultural and social factors be taken into account and used sustainably (Katajamäki, 2011).

The European Union cross-border cooperation program Interreg A supports cooperation between regions from at least two different member states lying directly on the borders or adjacent to them (EC, 2018). It is an instrument for regional development (EC, 2018) but at the same time, EU's cross-border cooperation has been seen as a tool for policy transfer from the EU to the local and regional level. This has also happened in the Arctic border region between Finland and Sweden (Jakola, 2016). "Policy transfer" refers to the process where knowledge related to policies, administrative arrangements and institutions in a certain time and location has been used to develop policies, administrative arrangements and institutions in another time and/or location (Dolowitz and Marsh, 1996, p. 344).

Medeiros (2010, pp. 442–443) notes that in the border regions of Sweden and Norway, and between Spain and Portugal, Interreg projects have played a major role in building more diffused and balanced regional development by creating and strengthening cross-border networking and contacts and by creating physical cross-border connections like roads. Still, Interreg projects have not been able to narrow the socioeconomic gap in these two regions when compared to the surrounding regions. According to Domínguez and Pires (2014), EU cross-border cooperation has brought people closer together on both sides of the border, and it has also promoted mutual interest and trust. The EU's cooperation structures play a major role in encouraging the actors in regional cooperation, but they should take a more individualized approach to cooperation in the regions. The structures should not be just tools to transfer EU guidelines to regions; they should become dynamic actors that improve the regional development (Domínguez and Pires, 2014, pp. 23–48). Still, according to Paasi and Prokkola (2008, p. 27), attitudes toward EU cross-border cooperation are positive in the European border regions, and the actors have started to recognize borders not as objects but as sources of regional development.

In the Arctic border region between Finland and Sweden, 43,450 people live on the Swedish side of the border, most of them in Kiruna (23,170 in 2017) (Statistics Sweden, 2018). Institutional cross-border cooperation started in this region in 1957 with the establishment of the North Calotte region. There have since been a number of cross-border cooperation agreements. Finland and Sweden joined the European Union in 1995, which provided new tools and funding opportunities for institutional cross-border cooperation and regional development in the region – not forgetting national and Nordic sources of finance. For example, EU cross-border cooperation and projects as operational instruments are now a part of everyday life in Finnish Tornio and Swedish Haparanda. The EU membership has made institutional cross-border cooperation a conscious strategy for regional development (Prokkola et al., 2015, p. 108).

The Arctic border region between Finland and Sweden is part of the Interreg A Nord program region (Map 7.2), which is one of the regional Interreg A programs. The goal in the current Interreg V A Nord program period (2014–2020) is to strengthen the competitiveness and attractiveness of the program area by developing the program region economically, socially and environmentally sustainably, and it aims at supporting cross-border cooperation in order to strengthen economic and social development. In the present Interreg V A Nord period, the budget is 83,217,500 euros (The Interreg Nord Programme, 2014, pp. 6, 55).

The Interreg V A Nord program region contains the northern parts of Finland, Sweden and Norway. The program region differs from other European regions due to its Arctic features with cold climates, polar nights, vast natural areas and abundance of natural resources. The region also holds the only indigenous people of the European Union, the Sámi people. In 2012, the population in the program region was 1,634,784. The biggest cities were Oulu (population 190,847), Umeå (117,294), Luleå (74,905), Skellefteå (71,774), Troms (69,116) and Rovaniemi (60,877) (Interreg Nord, 2014, p. 29).

The Interreg V A Nord program region is large and diverse. It includes municipalities such as Oulu, where the number of inhabitants is growing; Kemi and Luleå, which have a strong basic industry along with higher education, for example, in the technical field and Troms with very low unemployment. At the same time, in many rural parts of the program region, both unemployment and outmigration are high, and the population is aging fast (Interreg Nord, 2014).

In the Finnish regional development system today, the municipalities are strongly involved in the development of their territories; the development of local areas doesn't depend on just EU or national expertise and resources. Regional

Legend
- Cities
- Region Nord
- Region Sápmi

Vadsø

Karasjok

Tromsø

Inari

Kiruna

Bodø

Rovaniemi

Tornio

Luleå

Oulu

Snåsa

Umeå

Kokkola

Östersund

© The County Administrative Board of Norrbotten ©
National Land Survey of Sweden, The Geodata
Cooperation © National Land Survey of Finland
© Norwegian Mapping and Cadastre Authority and
© The Sami Parliament

0   50   100   200 Kilometers

Map 7.2 A map of the Interreg V A Nord program region (Interreg Nord, 2018)

development clearly benefits from local communities getting organized as it is not possible for only one body to "develop an area" (Sotarauta, 2015, p. 220).

There are 311 municipalities in Finland (in 2017). Finnish municipalities have self-government autonomy, which is secured in the Finnish Constitution and is based on representative democracy (Ijäs, 2017). According to the Local Government Act, municipalities aim at enhancing the well-being of their inhabitants and the vitality of the region, and they organize services for the inhabitants economically, socially and environmentally sustainably (Finlex, 410/2015, 1§).

Views on EU cross-border cooperation and the municipalities as research material

The research material is twofold, consisting in part of the description and SWOT analysis of the Interreg Nord program region (Interreg Nord, 2014), which is an appendix to the Joint Operational Program description of the Interreg V A Nord program (years 2014–2020) (2014). Data is also drawn from the municipal strategies of Tornio, Ylitornio, Pello, Kolari, Muonio and Enontekiö. The Joint Operational Program description lays out the content of the program as well as the basis for its operational activities. The description says that the program supports and encourages development of the program region based on the region's own strengths (principles in the selection of projects) (Hankevalikoiman periaatteet, 2014). This concurs with the current regional development idea that supporting a region's endogenous strength-based resources is the key to development.

In order to establish the potential of the Interreg Nord program area to grow, I examined the description and SWOT analysis to clarify the strengths, weaknesses, opportunities and threats in the program region of Interreg V A Nord (2014–2020) (Euroopan alueellinen yhteistyöohjelma, 2014, p. 5). I have further studied the strengths identified in the description and SWOT analysis, and these strengths have been understood as resources in the program region.

Also, all Finnish municipalities are required to have a strategy drawn up by the municipal council with long-term economic and other guidelines. The municipalities are led on the basis of their strategies, which rely on an estimation of the present situation and on future changes in the operational environment. The municipal strategies pay particular attention to promoting the welfare of the municipality, organizing and producing services, supporting the vitality of the living environment, etc. (Finlex, 410/2015). These strategies are the basis for any development work in the municipalities.

All six municipalities under study – Tornio, Ylitornio, Pello, Kolari, Muonio and Enontekiö – have drawn up strategies of their own together with several actors, including local inhabitants. The strategies are structured very differently and vary from 2-page overall vision-based documents to detailed documents covering 36 pages. As a whole, the strategies identify several endogenous strengths as resources for development, and these have been duly analyzed (Tornion kaupunki,

2018; Ylitornio, 2018; Pello, 2015; Kolarin kunta, 2017; Muonio, 2018; Enontekiön kunta, 2017).

Analyzing EU and municipality views

The research material has been systematically analyzed through content analysis, which is a way of organizing, describing and quantifying phenomena. In content analysis, the aim is to create models that show the phenomena in a compressed form according to which the phenomena can be conceptualized (Kyngäs and Vanhanen, 1999, p. 2). In the following, the local resources referred to in the research material will be examined according to four aspects and categories – environmental, economic, cultural and social resources – as the basis for regional development. The aspects overlap as the resources can be a part of several categories. For example, nature can be recognized as a resource in the environmental, economic and cultural categories.

Environment-related issues were mentioned as resources in the strategies in five of the six municipalities: Ylitornio (2018), Pello (2015), Kolari (2017), Muonio (2018) and Enontekiö (2017). Nature emerged both in terms of natural values and as a part of everyday life. The diversity of nature and the beautiful landscape were clearly appreciated, as was the possibility for inhabitants and visitors to enjoy recreation in a natural environment and to pursue nature-related hobbies (Ylitornio, 2018; Pello, 2015, p. 7, appendix; Kolarin kunta, 2017; Muonio, 2018; Enontekiön kunta, 2017). Other strengths included the possibility of building a house in a beautiful landscape (Kolarin kunta, 2017) and overall security, peace and quiet and a clean living environment (Ylitornio, 2018; Pello, 2015, p. 7, appendix). Also, the living environment was felt to be of high quality yet not expensive (Pello, 2015, p. 7, appendix). And while the municipality of Enontekiö saw nature as a source of livelihood, they also appreciated the value of nature as such: nature was recognized as a main value for the inhabitants (Enontekiön kunta, 2017). Nature and a nature-related living environment were highly appreciated as integral parts of everyday life.

The description and SWOT analysis of the Interreg Nord program region also identified nature as a resource. Nature values were emphasized, not least because nature in the region is unique and clean with plenty of fish, wild animals, flora, national parks and protected areas. The diversity of nature was also mentioned as an opportunity for the region's inhabitants and visitors to pursue nature-related experiences and activities.

Nature was similarly seen as important for preserving the cultural heritage of the Sámi people. In fact, nature, culture and different kinds of recreational resources may enhance the appeal of the region when people consider moving to or staying there (Interreg Nord, 2014, pp. 3–4, 23).

Examining nature as a (non-business) resource in the data sources shows that nature is an important and diverse resource not only in the everyday lives of the local inhabitants but also as a source for experiences and activities for visitors. Still, the municipal strategies focus on this aspect much more clearly than the description and SWOT analysis of the Interreg Nord program region, which does recognize nature as a resource but in a wider perspective, further away from the everyday basis.

What the description and SWOT analysis clearly emphasize are nature and its diversity as economic resources and in terms of employment (Interreg Nord, 2014, p. 19). The program region has a unique nature and special environmental conditions along with growing purchasing power and cross-border cooperation – obvious benefits for environmentally and culturally sustainable tourism. Further strengths include the traditional sources of livelihood of the Sámi people with long-term ecological sustainability, as well as their business opportunities in experience-based tourism and in the culture, food and service sectors (Interreg Nord, 2014, pp. 9, 21).

Nature-based growth in, for example, bioenergy and environmental technology also counts as a strength (Interreg Nord, 2014, p. 28), as do Arctic environmental conditions together with high know-how and high technology in terms of the testing and practicing field, for example, in the car industry (Interreg Nord, 2014, pp. 9, 21).

Other industries identified as economic resources in the description and SWOT analysis of the Interreg Nord program region ranged from strong basic industries (ore and minerals, oil, gas, forestry and fishing) and large raw material–based industries to mining and steel industries as well as metal and chemical industries and communication technology. Sources of raw materials and energy were estimated to be large. Here, too, the importance of the culturally and environmentally sustainable use of resources was emphasized (Interreg Nord, 2014). The role of companies willing to invest in the region emerged as crucial. This has a positive effect on the program region as a whole. That there is a concentration of people in some areas was seen as a benefit: the region benefits from a "critical mass" in testing innovations and expanding their use (Interreg Nord, 2014, pp. 6, 19, 22, 28).

The importance of education, innovation and research is obvious. Growing entrepreneurial knowledge along with strong environments of education as well as research and innovation – and good access to them – were seen as valuable resources (Description and SWOT analysis of Interreg Nord program region, 2014, p. 28). The active R & D work in several locations has produced special skills within Arctic technology, space technology, biotechnology, marine fishing, marine biotechnology, marine engineering, mining and mineral technology (Description and SWOT analysis of Interreg Nord program region 2014, pp. 10, 11, 14, 23).

Businesses or sources of livelihood also appeared as a resource in the municipal strategies, and often these resources were related to nature. The strategies found that the structures of livelihood were diverse, including agriculture and forestry, traditional reindeer herding, use of natural products, local food trade, mining industry, know-how in bioenergy industry; building, logistics and small-scale industries were also mentioned as resources (Ylitornio, 2018; Pello, 2015, p. 7, appendix; Kolarin kunta, 2017). Tourism earned a place of its own in the strategies of Ylitornio (2018), Pello (2015, appendix) and Kolari (2017). Kolari identified itself as the center of nature tourism, with a long tradition in tourist services both in Kolari and the most popular national park, Pallas-Yllästunturi (Kolarin kunta, 2017). Pello (2015) takes pride in having the best salmon river in Europe and possibilities for fishing tourism (2015). The municipal strategy of Kolari (2017) also mentions municipal optical fiber as a resource.

Ylitornio finds that entrepreneurs benefit significantly from the population in Övertorneå on the Swedish side of the border (Ylitornio, 2018). Similarly, the municipal strategy of Pello appreciates the joint markets across the border and cross-border cooperation (Pello, 2015, p. 7). The population on the other side of the border translates into a larger customer base.

Comparison of the economic resources cited in the two data sources shows that the recognized resources differ to some extent. While tourism and industry were included in both document types, fields related to, for example, Arctic technology, space technology, marine fishing, marine biotechnology and marine engineering featured only in the Interreg Nord program region document. In all likelihood, they were omitted from the municipal strategies because they are not relevant in these municipalities. This also shows how large and diverse the program region is with different kinds of developmental possibilities and needs.

The border and border crossing were identified as economic resources in both data sources. They were seen as resources not only for tourism, but also for business in general, thanks to the larger customer base on the other side of the border.

Cultural resources were highlighted in the strategies of Pello (2015), Kolari (2017) and Enontekiö (2017). Enontekiö, for example, emphasized its location in the Sámi region and by three national borders; the cultural environment is diverse (Enontekiön kunta, 2017). Pello took pride in its unique local culture, stories and language (Pello, 2015, p. 7, appendix), and handicrafts expertise was mentioned as a resource in the strategy of Kolari (2017).

The description and SWOT analysis of the Interreg Nord program region recognized as resources the rich cultural heritage and diversity in languages and cultural traditions. The region is home to several nationalities, cultures, languages and ways of living, and the centuries-long interaction has enriched the area. People are used to crossing national borders; in the border regions, in particular, relations to neighbors have been a natural part of everyday life, based partly on a shared history. The cultural and natural heritage within the region makes the program area unique and attractive.

As a resource, culture inspires the development of new services, industries and attractions and also drives local people's creativity and creates local and regional cohesion. That Sámi people reside across the four countries with common languages, cultures and traditional industries was clearly identified as a resource. The Sámi cultures and social life enrich the countries' collective culture and social life (Description and SWOT analysis, 2014, pp. 16, 26, 28).

Culture was a recognized resource in both data sources. The municipal strategies were less specific about the ways in which culture constitutes a resource, but the description and SWOT analysis made the case that culture was important for business as well as for local creativity and regional cohesion.

The program region description and SWOT analysis also saw that the diversity of nationalities, communities and cultures, along with border crossing as a part of everyday life amounted to being resources. This is how the border itself can be recognized as a cultural resource.

While cultural resources can also count as social resources, the research material obviously acknowledges other social resources too. Local inhabitants are a key social resource, identified in the municipal strategies of Ylitornio, Pello, Kolari and Enontekiö. The inhabitants were portrayed as working actively in various associations and organizing events, and the third sector was commended. The social resource aspect also covers a sense of community: the inhabitants show an interest in their environs and happily drop by their neighbors (Ylitornio, 2018; Pello, 2015; Kolarin kunta, 2017). The easygoing nature of the inhabitants and the fact that people know each other were greatly appreciated (Enontekiön kunta, 2017). The strategy of Kolari found that the large families (compared to other municipalities in Lapland) were also a resource, as were the 16 local villages, each with its own characteristics (Kolarin kunta, 2017).

Ylitornio (2018), Pello (2015) and Kolari (2017) praised the well-functioning and high-quality municipal services – clearly a resource. Ylitornio found that it has a good image and a solid working and decision-making atmosphere. The municipal personnel are well educated and motivated, and there are mostly enough personnel to provide the services. Ylitornio has a robust financial situation, which enables further educating personnel and providing modern facilities and working premises (Ylitornio, 2018).

Ylitornio (2018) also appreciates its central location. There are possibilities for cooperation across the municipal borders but also across the national border between Finland and Sweden. This argument is most likely related to co-organizing some municipal services with partners on the other side of the border. Location as a resource also came up in the strategies of Pello (2015), Enontekiö (2017) and Ylitornio (2018). This refers to being located by the national border, at the crossroads of three states, in the Arctic Circle, in the North Calotte and by the Tornio River and being related to neighboring municipalities (Pello, 2015; Enontekiön kunta, 2017).

Location appeared most prominently in the municipal strategy of Tornio. Finnish Tornio and Swedish Haparanda are twin cities with a long history of active cooperation, for example, in organizing public services. The gist of Tornio's municipal strategy is to be found in the title of the strategy, translated as "The most functioning border city 2027." In this, location by the national border between Finland and Sweden was *the* resource (Tornion kaupunki, 2018).

The description and SWOT analysis of the Interreg Nord program region identified Sámi people, Sámi culture and Sámi societal life as resources in the program region. Another human resource is the local people in concentrated areas as a critical mass for the business sector. The concentrated areas also seem to have working infrastructures. Other strengths included active and long-established cross-border interaction. Having access to means of transport – airports, railroads, borders and knowledge-intensive traffic – is clearly important in these northern locations (Description and SWOT analysis of Interreg Nord program region, 2014, p. 15).

Comparing the social resources in the research material again shows both similarities and differences. The municipal strategies focus on local, active people and a sense of community as significant resources in people's everyday lives. The description and SWOT analysis of the Interreg Nord program region highlights Sámi people as a social resource in relation to their cultural and societal life, and people in concentrated areas as a business resource.

Well-functioning services and a good decision-making atmosphere on the local level were appreciated as a resource in the municipal strategies of Ylitornio, Pello and Kolari while the description and SWOT analysis drew attention to a solid infrastructure in the concentrated areas. Ylitornio, Pello and Kolari are not located in these concentrated areas but argued all the same that their municipal services are in good working order. This did not come up in the description and SWOT analysis. It seems that the municipalities see their social resources as being more diverse and abundant than one would find on the basis of the description and SWOT analysis alone.

It is surprising for the municipality of Ylitornio to argue that it has a central location. One might contend that the location is peripheral, far from national centers. Still, as the border between Finland and Sweden is genuinely open, it is an open door to other communities that, in fact, are closer than one might think. Borders can be seen as lasting borderlines but also as practices and discourses that spread to the whole society (Paasi, 1999, p. 670), and this is visible on the Arctic border between Finland and Sweden. The border does not exist as an autonomous actor but through the meanings attached to it (Sohn, 2015, pp. 2, 5): borders have different meanings to different people (Balibar, 2002, p. 81).

Conclusion

Regional development activity is now directed especially at strengthening a region's endogenic potential to develop its own resources (Sotarauta, 2015, p. 215). As Katajamäki (2011) points out, regional development today requires that environmental, economic, cultural and social factors be taken into account. This is evident both in the municipalities' development strategies and in the EU's cross-border cooperation program Interreg V A Nord, which both aim at promoting regional development in the Arctic border region between Finland and Sweden. Only the emphasis is different.

Resources recognized in the municipal strategies pertain to good everyday life. What is appreciated are a clean, safe and peaceful living environment in or very near nature; good services; a sense of community and social life, as well as cultural aspects together with sources of livelihood. This is a local-level-oriented picture based on the needs of living a good life.

The description and SWOT analysis of the Interreg Nord program region emphasize the economic resources. Environmental, cultural and social resources are also appreciated, but they appear to be driven by business considerations.

The picture that emerges from the Interreg V A Nord research material is one focusing on the wider level rather than the local everyday concerns. This is understandable in light of the fact that the program region is large and diverse with different needs and potentials.

The crosscutting resource mentioned in both data sources is location, which emerged as a resource related to economic, cultural and social resources. Location in the Arctic region is an asset, for example, for tourism, and being located close to the open national border offers possibilities for business, cross-border culture and organizing services together. While location appeared as a resource in both data sources, it featured more prominently in the municipal strategies. The strategy of Ylitornio went as far as seeing its location as central – in relation to neighboring municipalities and cities, as well as to Sweden, the neighboring country. This emphasizes the fact that borders are now recognized as a resource for regional development (Paasi and Prokkola, 2008, p. 27).

Municipalities and EU cross-border cooperation are different kinds of tools for regional development, and this is visible in the research material. Municipalities concentrate on improving the everyday lives of their inhabitants, whereas EU cross-border cooperation focuses more on issues such as employment and cohesion within the EU. These differences are visible also in the way the two actors recognize the local resources and identify the key resources in the development of the region.

Because the program region of the Interreg V A Nord varies a lot, the resources of the program region vary as well. This can be challenging for regional development and for choosing the right resources to strengthen in order to increase and support development in the program region. Diverse resources are also related to push-and-pull factors when people are considering where they want to live. At the same time, the diverse resources in the program region can be a positive aspect when there are certain resources close by for the needs of other parts of the program region that are lacking them.

The environmental, economic, cultural and social aspects in regional development overlap and interlace, and they require the functioning of each other. For example, without a clean, safe and well-functioning living environment that provides public services, cultural facets and sources of livelihood, it is difficult for social resources – say, local inhabitants – to exist. At times, however, it can be hard to sustainably combine the use of different resources. For example, how to combine the use of different resources such as traditional sources of livelihood, tourism and the use of natural resources so that the use of one resource does not harm the other, and they all can be used sustainably? Also, in the Arctic border region between Finland and Sweden, the resources are diverse, and combining their use is sometimes challenging. For example, how to fit wind turbines into the landscape and consider the needs of reindeer herding, local inhabitants and tourism?

Regional development must not be shortsighted. Rather, it has to be done sustainably so that resources can also be utilized in the future. The importance of supporting environmentally, economically and socially sustainable development was emphasized both in the Interreg A Nord program and in the Local Government Act of the Finnish municipalities. Cultural development has probably been considered as part of the social aspects of development. On the Finnish side of the Finnish-Swedish border region, the most urgent and crucial question for the development of the region is how to maintain social resources sustainably when the population is aging and out-migration is heavy. Additionally, cultural resources such as the shared language across the national border are partly based on shared history, creating cross-border cohesion, but will this joint cultural resource be alive in the future? What happens when the time of a united region without national borders goes further down in history; how can this cultural resource survive? Younger generations on the Swedish side of the border already speak Finnish, or so-called *meänkieli*, less well than the older generations. Still, there are actors like the Meänmaa association, which works for maintaining the cross-border cultural heritage. Furthermore, some natural resources such as ore are not sustainable resources, even if they do last long into the future.

The research region might be peripheral, far from the national and European core centers, but it still has its own strengths. According to the research material, the Arctic border region on the Finnish side between Finland and Sweden has many endogenic resources to enable the living of a good life and the sustainable development of the region. Are these resources enough, and have the resources been used in the best possible way in order to promote the development of the region? Is the region lacking some crucial resources in supporting the regional development, and if so, how could this be rectified? Furthermore, will these also be the right resources for living a good life in the region in the future, and what kind of life will be good in the future?

The impact of exogenous factors such as global warming obviously influences the regional development globally, and different kind of resources may be needed in the future. In regional development, the future prospects and the needs they bring have to be evaluated as well as possible. No one knows for sure what the future holds, but when the resources of a region are diverse, preparing for future needs might be a bit easier.

References

Balibar, É. (2002) *Politics and the other scene*. New York: Verso.

Dolowitz, D. and Marsh, D. (1996) "Who learns what from whom: a review of the policy transfer literature," *Political Studies*, 44, pp. 343–357.

Domínguez, L. and Pires, I. (2014) "EU cross-border cooperation: historical balance and future perspectives" in Domínguez, L. and Pires, I. (eds.) *Cross-border cooperation structures in Europe: learning from the past, looking to the future*. Euroclio No. 82. Brussels: P.I.E. Peter Lang, pp. 23–47.

Enontekiön kunta (2017) *Son rohki soma paikka ellää. Enontekiö. Met tehemä yhessä. Mii bargat outtas. Strategia 2025* [Online]. Available at: https://enontekio.fi/web2017/wp-content/uploads/2018/02/Met-tehemä-yhessä-strategia-2025-kv-20.12.2017.pdf (Accessed: April 25, 2019).

Euroopan *alueellinen yhteistyöohjelma. Pohjoinen 2014–2020. Hallituksen päätöksen N2013/5423 RT liite* (2014) [Online]. Available at: www.interregnord.com/wp-content/uploads/Interreg-Pohjoinen-FIN-web.pdf (Accessed: April 25, 2019).

European Commission (2018) *Interreg A—Cross-border cooperation* [Online]. Available at: https://ec.europa.eu/regional_policy/en/policy/cooperation/european-territorial/cross-border/ (Accessed: April 25, 2019).

Finlex (2015) *Kuntalaki 410/2015* [Online]. Available at: www.finlex.fi/fi/laki/alkup/2015/20150410 (Accessed: April 25, 2019).

Hankevalikoiman periaatteet [Principles in the selection of projects] (2014) In *Pohjoinen 2014–2020. Euroopan alueellinen yhteistyöohjelma* [Online]. Available at: www.interreg nord.com/wp-content/uploads/Interreg-Pohjoinen-FIN-web.pdf (Accessed: January 24, 2019).

Ijäs, S. (2017) *Uusi kuntalaki pintaa syvemmältä.* Kuntaliitto. Ylivieska and Oulu: The Summer University of Northern Ostrobothnia.

Interreg Nord (2014) "Områdesbeskrivning och SWOT-analys av Interreg Nords Programområde" in *Samarbetsprogram inom målet Europeiskt territoriellt samarbete. Officiella Nordprogrammet inkl bilagor* [Online]. Available at: www.interregnord.com/wp-content/uploads/2015/01/Officiella-Nordprogrammet-inkl-bilagor-12-dec-2014.pdf (Accessed: January 18, 2019).

Interreg Nord (2018) *Programme region* [Online]. Available at: www.interregnord.com/wp-content/uploads/lstbd_interreg_swe_300dpi.jpg (Accessed: April 25, 2019).

The Interreg Nord Programme 2014–2020 (2014) *Borderless opportunities* [Online]. Available at: www.interregnord.com/wp-content/uploads/interreg_popvers_eng-korrad-160504.pdf (Accessed: April 25, 2019).

Jakola, F. (2016) "Borders, planning and policy transfer: historical transformation of development discourses in the Finnish Torne Valley," *European Planning Studies*, 24, pp. 1806–1824. doi:10.1080/09654313.2016.1194808

Jańczak, J. (2018) "Integration de-scaled: symbolic manifestations of cross-border and European integration in border twin towns," *Journal of Borderlands Studies*, 33(3), pp. 393–413.

Katajamäki, H. (2011) *Mitä on aluekehittäminen?* [Online]. Available at: www.uva.fi/fi/blogs/expert/aluekehityksen_arki/mita_on_aluekehittaminen/ (Accessed: January 26, 2017).

Kolarin kunta (2017) *Kolarin kolme kovaa tavoitetta* [Online]. Available at: www.kolari.fi/fi/tietoa-kunnasta/kuntastrategia.html (Accessed: April 25, 2019).

Kyngäs, H. and Vanhanen, L. (1999) "Sisällön analyysi," *Hoitotiede*, 11(1), pp. 3–12.

Medeiros, E. (2010) "Old vs recent cross-border cooperation: Portugal-Spain and Norway-Sweden," *Area*, 42(4), pp. 434–443.

Muonio (2018) *Kuntastrategia 2025* [Online]. Available at: www.muonio.fi/media/ahal linto/strategiat/muonio_kuntastrategia_luonnos_21.5.pdf (Accessed: April 25, 2019).

Paasi, A. (1999) "Boundaries as social practice and discourse: the Finnish-Russian border," *Regional Studies*, 33(7), pp. 669–680.

Paasi, A. and Prokkola, E-K. (2008) "Territorial dynamics, cross-border work and everyday life in the Finnish-Swedish border area," *Space and Polity*, 12(1), pp. 13–29.

Pello (2015) *Pello. Kuntastrategia 2015–2020* [Online]. Available at: www.pello.fi/pello/wp-content/uploads/2015/10/pello_kuntastrategia_28092015.pdf (Accessed: April 24, 2019).

Pike, A., Rodriguez-Pose, A. and Tomaney, J. (2006) *Regional and local development.* Abingdon: Routledge.

Prokkola, E-K., Zimmerbauer, K. and Jakola, F. (2015) "Performance of regional identity in the implementation of European cross-border initiatives," *European Urban and Regional Studies*, 22(1), pp. 104–117.

Ruotsala, H. (2011) "Kaksi kukkaroa ja kaksi kelloa. Ylirajaisuutta ja monipaikkaisuutta Tornio-Haaparannan kaksoiskaupungissa," *Sananjalka*, 53, pp. 196–217.

Sohn, C. (2015) *On borders' multiplicity: a perspective from assemblage theory*. Working Paper 10. EUBORDERSCAPES.

Sotarauta, M. (2015) "Aluekehittämisen kehityskaari Suomessa ja peruskäsitteet" in Karppi, I. (ed.) *Governance—hallinnan uusia ulottuvuuksia*. Tampere: Tampereen yliopisto, pp. 215–230.

Statistics Finland (2018a) *StatFin: labour market* [Online]. Available at: https://pxnet2.stat.fi/PXWeb/pxweb/fi/StatFin/#_ga=2.83788630.154213562.1547992160-1286175287.1515162649 (Accessed: April 24, 2019).

Statistics Finland (2018b) *StatFin: population* [Online]. Available at: https://pxnet2.stat.fi/PXWeb/pxweb/fi/StatFin/#_ga=2.83788630.154213562.1547992160-1286175287.1515162649 (Accessed: April 24, 2019).

Statistics Sweden (2018) [Online]. Available at: www.scb.se/en/ (Accessed: April 24, 2019).

Tödtling Franz (2011) "Endogenous approaches to local and regional development policy" in Pike, A., Rodríguez-Pose, A. and Tomaney, J. (eds.) *Handbook of local and regional development*. Abingdon and New York: Routledge, pp. 333–343.

Tornion kaupunki (2018) *Maailman toimivin rajakaupunki 2027. Tornion kaupunkistrategia 2017–2021* [Online]. Available at: https://indd.adobe.com/view/092a56da-1a0c-4956-a0ff-2dd5d95d23ee (Accessed: April 24, 2019).

Ylitornio (2018) *Ylitornion kuntastrategia Meän Väylä* [Online]. Available at: https://ylitornio.fi/wp-content/uploads/2018/05/Ylitornio-kuntastrategia.pdf (Accessed: April 24, 2019).

Part II
Whose imaginaries?

8 The political ecology of Northern adaptation

Power, nature and knowledge

Gemma Holt

Introduction: why adaptation?

There is widespread scientific consensus that climate change is happening and that human activities are the primary driver (IPCC, 2014, p. 2). Reducing greenhouse gas emissions is of the utmost urgency; however, measures to reduce emissions are only one piece of the climate change challenge. Whether they take the form of rising temperatures, greater incidences of extreme weather events or pressure on water resources, the impacts of climate change will become increasingly severe (IPCC, 2014, p. 8). Responding to these consequences is as much a social science issue as it is a natural science issue. Adaptation was defined in the Intergovernmental Panel on Climate Change's (IPCC) Fifth Assessment Report as

> the process of adjustment to actual or expected climate and its effects. In human systems, adaptation seeks to moderate or avoid harm or exploit beneficial opportunities. In some natural systems, human intervention may facilitate adjustment to expected climate and its effects.
>
> (IPCC, 2014, p. 118)

Adaptation is often contrasted with mitigation, which the report defines as "a human intervention to reduce the sources or enhance the sinks of greenhouse gases" (IPCC, 2014, p. 125). In other words, mitigation is avoiding climate change, whereas adaptation is coping with it.

How does adaptation relate to sustainability?

As other chapters in this volume contend, "sustainability" implies the maintenance of an existing way of life. Sustainability focuses on the preservation of a status quo, which by definition is fundamentally at odds with the notion of adaptation. What I wish to highlight in this chapter is that how we adapt to climate change has implications for the social and cultural dimensions of sustainability. Adaptation is not an inherently sustainable process; in fact, multiple authors have explored the possibility that adaptation may be detrimental to vulnerable communities (Loring et al., 2016) or the environment (Staudt et al., 2013).

Ensuring social sustainability requires that adaptation not just be about finding technical solutions to specific climate impacts but rather about designing adaptation processes that encompass social and climate justice (Schlosberg et al., 2017, p. 413).

Conventional approaches to understanding climate change have focused on identifying and quantifying long-term climate impacts on different ecosystems. However, this top-down approach has largely failed to consider regional and local impacts and the capacity of communities to adapt to climate-induced changes (Granderson, 2014, p. 55). The alternative is to situate adaptation actions in a social context; consequently, effective adaptation strategies integrate climate science with an understanding of the needs and values of human society. These strategies have to take into account different forms of knowledge, a variety of stakeholders and an understanding of local conditions and decision-making processes (Ford et al., 2015, p. 176).

Knowledge has played a critical role in the evolution of the Arctic

Adaptation strategies are of critical concern in the Arctic, which is warming at twice the rate of the rest of the globe (NOAA, 2018). This warming is expected to have significant impacts on Arctic ecosystems and economies. Addressing this issue is a priority at the highest level of Arctic governance, where the primary forum for debate is the Arctic Council.

The Arctic Council is a high-level intergovernmental forum established in 1996 to promote cooperation and coordination on issues of sustainable development and environmental protection. The Arctic Council consists of eight Arctic states: Canada, Denmark, Finland, Iceland, Norway, Sweden, Russia and the United States; six permanent participants, all of which are regional indigenous peoples' associations and over thirty observers, which include non-Arctic states, intergovernmental and interparliamentary organizations and nongovernmental organizations (Arctic Council, 2018).

Predating the creation of the Arctic Council, science played a significant role as a determinant of public policy agendas and institutions in the region. In 1987, during an era of softening East-West tensions, Soviet president Mikhail Gorbačëv Mihail outlined a future where "the community and interrelationship of the interests of our entire world is felt in the northern part of the globe, in the Arctic, perhaps more than anywhere else" (Gorbačëv, 1987). Gorbačëv recognized that "scientific exploration of the Arctic is of immense importance for the whole of mankind" and called for the creation of a "joint Arctic research council" (1987).

Gorbačëv speech signified a turning point for Arctic cooperation. Three years later, the International Arctic Science Committee was founded to coordinate scientific activity in the polar region. The following year, the eight Arctic states signed the Arctic Environmental Protection Strategy, which outlined a "common future" for the region and addressed monitoring, assessment, protection, emergency preparedness/response and conservation (Arctic Council, 2011a, para. 5).

The process of formalizing the Arctic science regime culminated in the establishment of the Arctic Council in 1996. Since then, the council has played an important role as a policy-shaping institution, primarily through the outputs of its six scientific working groups. These groups have produced assessments on topics ranging from persistent organic pollutants and black carbon mitigation to marine use and human health. Collectively, the science-based assessments and related policy documents comprise a singularly comprehensive base of knowledge about the region (Brigham et al., 2016, para. 7).

Adaptation Actions for a Changing Arctic is an example of the developing Arctic science regime

The Arctic Monitoring and Assessment Programme, more frequently referred to as AMAP, is one of the six working groups of the Arctic Council. AMAP has three primary tasks: monitoring and assessing issues relating to climate change and pollution, documenting trends and their effects on ecosystems and humans and producing science-based, policy-relevant assessments to inform policy and decision-making processes at the government level (AMAP, 2019). As the impacts of climate change in the Arctic have accelerated, AMAP's research has focused on the interrelated and interlinked drivers of change as well as the human and environmental impacts. Projects like Adaptation Actions for a Changing Arctic, which was led by AMAP with contributions from the other Arctic Council working groups, permanent participants to the Arctic Council, stakeholders in the three pilot regions and international organizations, demonstrate how AMAP's research has evolved to fill gaps in the knowledge base.

The AACA is the result of a 2011 directive to "review the need for an integrated assessment of multiple drivers of Arctic change as a tool for Indigenous peoples, Arctic residents, governments and industry to prepare for the future" (Arctic Council, 2011b, p. 4). Such broad objectives required a diverse group of researchers and authors. The peer-reviewed Barents report had over one hundred contributing authors. The diversity of authorship is reflective of the region as a whole: the Barents area is home to 5 million people in four countries. Its inhabitants include Swedes, Finns, Norwegians, Russians and a number of indigenous peoples. The region is historically rural with an average population density of 2.9 inhabitants per square kilometer, but the population is generally becoming more urbanized (AMAP, 2017a, p. 22). Demographic research is emphasized in the AACA as a measurement of key socioeconomic changes in the region.

The content of the report is broad in scope, addressing both environmental and social changes in the Barents region. Its chapters closely examine how multiple interconnected factors associated with climate change are affecting local communities and consider potential future pathways. The AACA report explains that

> adaptation is both a new policy field and a normal part of everyday life for individuals, communities, corporate actors, and whole societies as they adjust their activities in relation to observed and anticipated changes.
>
> (AMAP, 2017a, p. xii)

Adaptation in the Arctic combines climatic factors with demographic, economic, technological and political drivers. It is proactive and planned management of change that leads to new socioecological configurations capable of functioning under new conditions.

In addition to the full scientific report, AMAP also released a summary for policymakers, which is standard practice for the working groups of the Arctic Council. These reports are intended to be short, readable syntheses of longer, more technical research documents. Summary reports are consistent with the project's goal of enabling "more informed, timely and responsive policy and decision-making related to adaptation actions in a rapidly changing Arctic" (AMAP, 2014, p. 2). To that end, the AACA report and other assessment reports produced by Arctic Council working groups are explicitly designed to convey usable knowledge.

The full AACA report illuminates how knowledge functions as a resource

The AACA shifts the focus on adaptation toward societies and economies, whereas most research situates adaptation within analyses of climate. The report explicitly demonstrates how biophysical changes have socioeconomic consequences. The authors examine the cumulative impacts of and responses to change in a wide range of sectors in both indigenous and non-indigenous communities. They juxtapose how climate change will impact different economic sectors with possible adaptive actions and strategies. For example, the agricultural sector will be impacted by a longer growing season, shorter winters and earlier spring, increased precipitation in autumn and summer and changing snow and ice cover. In turn, this could lead to increased yields, damaged harvests, better growth conditions for trees, changes in crop variety and increased pests and diseases (AMAP, 2017a, p. 225).

In order to apply such a broad lens to the assessment of potential adaptation scenarios, a diversity of knowledge was required. The authors of the AACA represented different countries, institutions, disciplines and backgrounds. The report builds on a history of integrating multiple knowledges in Arctic environmental assessments, which has been an explicit focus of the Arctic Council and its working groups since the publication of the *Arctic Climate Impact Assessment* in 2005.

In particular, these assessments have evolved the process for integrating traditional ecological knowledge and scientific research. Traditional ecological knowledge is part of a complex political arrangement through which indigenous peoples gain "a seat at the table" (Nadasdy, 1999, p. 3). Their expertise functions as political currency and increases the likelihood that local environmental issues are discussed as a component of international environmental discourse. And since the 1990s, both Arctic and non-Arctic states have leveraged their scientific knowledge in order to shape policy. In this sense, knowledge in multiple forms functions as a resource in the Arctic, albeit in different capacities.

Multiple adaptation knowledges require translation

Knowledge today is understood as being multiple, with multiple claims to represent reality (Nyseth and Viken, 2014, p. 1). Scientific knowledge, as it is embedded in scientific institutions, is one of several options that can be used to inform planning and policy decisions. However, those institutions are one reason for the disconnect between the science and policy worlds. Scientists answer to scientific norms, research agendas, methodological standards and funders; policymakers answer to constituencies, political agreements and stakeholders (Royal Society, 2010, p. 16). Thus, there exists a gap in communication between climate researchers and climate policymakers.

Narrowing that gap is critical for successful adaptation. There have been consistent efforts to make climate science clearer and more relevant to policymakers, which is an important first step (Clark, 2016, p. 4571). However, that is not the only challenge in constructing coherent and effective adaptation policy. There are also the twin issues of increasing the inclusion of local knowledge and of moving debates about how to address climate change away from the exclusive control of "experts" and into the public sphere.

Collaborative scientific assessments like the AACA are the foundation of cooperation in the Arctic (Koivurova, 2010, pp. 146–147). To be effective, these reports have to balance the agendas of scientific researchers, indigenous experts and policymakers. Arctic scientific assessments influence how regional and international stakeholders perceive the region. They are echoed by news media, synthesized into summaries for policymakers and reproduced by scholars and researchers around the world. Because of the effect these reports have on perceptions of the region, the stakes for what knowledge is included are incredibly high.

Without an explicit mandate to incorporate traditional ecological knowledge, alternative ways of knowing are at risk of marginalization (Davis and Wagner, 2003; Barnhardt and Kawagley, 2005; Nadasdy, 1999). Of course, the Arctic Council *does* have a mandate to involve indigenous peoples in their roles as permanent participants. Indigenous peoples contribute to scientific reports, and indigenous experts work in tandem with western scientists to document the damaging effects of environmental change on traditional livelihoods. The AACA devotes an entire chapter to indigenous peoples' perspectives, assessing the trends on the economy, climate change and governance that are forcing the region's indigenous peoples to adapt.

However, the literature is critical of whether traditional ecological knowledge has ever truly been incorporated into Arctic science assessments (Nadasdy, 1999, p. 2). The contributions of indigenous experts are expected to supplement Western knowledge, which is more readily incorporated into policy. Traditional ecological knowledge is only legitimized when it is produced alongside Western science; otherwise, it is largely excluded from the discussion. Knowledge is both produced and utilized by those in power; the dominant discourse surrounding the politics of climate change privileges certain ways of framing the problem and developing potential solutions.

How, then, can scientific knowledge and social values be combined into policy processes? Whose knowledge is relevant for adaptation, and how can it be communicated effectively and linked to policy and practice? It is difficult to determine what constitutes policy-relevant knowledge when operating in uncertain, contested and complex knowledge landscapes (Moran et al., 2016, p. 132). However, adhering to the status quo constrains the spectrum of sustainable adaptation outcomes. Despite the perception that the Arctic is "a region in change," the practices by which laws and policies are produced have remained largely stagnant (Koivurova, 2010, p. 153). The dialogue around adaptation continues to evolve, but it is outpaced by the changes in the environment. Policymakers and researchers need a better understanding of the conditions that make research relevant in order to shorten the lag time between knowledge creation and implementation. It is necessary to reconfigure the mechanisms by which knowledge about the environment is produced, then used to construct policies and then labeled as legitimate in political debate.

The theoretical basis for making knowledge relevant to policy

Scientific knowledge about environmental change is shaped by social and political factors. As Jasanoff notes,

> scientific knowledge . . . is not a transcendent mirror of reality. It both embeds and is embedded in social practices, identities, norms, conventions, discourses, instruments, and institutions – in short, all the building blocks of what we term the *social*.
>
> (2004, pp. 2–3)

Scientific reports and scientific institutions – including universities, government agencies, museums and research organizations – reflect both social and political norms.

The assumption that environmental science can be separated from environmental politics is problematic; neither the creation of the science nor the policy options that result from it is politically neutral. Adaptation in particular is political "all the way through" (Eriksen et al., 2015, p. 1). Political processes determine whose perspective on adaptation is valued and whose adaptation goals are prioritized. Adaptation in any context always creates winners and losers, and what counts as adaptive is always contested. Thus, effective adaptation research must engage with multiple adaptation knowledges and stress how adaptation is situated within both social and environmental practices.

To illuminate the conditions that make research relevant in the formation of sustainable development policy, it is necessary to understand the underlying causes of environmental change and the process of decision-making simultaneously. One way to look at this is through the concept of "usable knowledge," the idea of which has been discussed in a variety of literature (e.g. Haas, 2016; Clark, 2016; Young, 2002). Peter Haas summarizes the concept:

> In short, usable knowledge encompasses a substantive core that makes it usable for policy makers, and a procedural dimension that provides a

mechanism for transmitting knowledge from the scientific community to the policy world."

(2004, p. 573)

The "substantive core" of usable knowledge has three characteristics: credibility, legitimacy and saliency (Haas, 2004). Researchers seeking to generate usable knowledge must consider all these features. Knowledge must first be *credible*: users must perceive the source as trustworthy, and they must assume that there is truth behind new discoveries and technologies. It must also be *legitimate*, meaning that it is regarded as free from bias and produced in a manner that can be replicated and understood. Finally, it must be *salient* and relevant to the needs of users.

Understanding what makes knowledge usable in adaptation literature is the first step in crafting policy-relevant research. In this context, research for research's sake will not generate substantive change. The worlds of science and policy often appear to be mutually exclusive and unreceptive to each other, even if they are attempting to solve the same problem through parallel efforts. The question is not "Does knowledge matter?" but rather "What type of knowledge matters?"

In a Northern context, this question is fundamentally political. The primacy of science in the Arctic Council means that whose knowledge matters determines who gets to participate. Observers to the Arctic Council leverage their scientific expertise to contribute to working groups; even if they are geographically distant from the Arctic, their knowledge makes them relevant (Koivurova, 2010, p. 150). These dynamics are further complicated by the differences between Western science and traditional ecological knowledge, which have prompted intense debate over what constitutes research (or science).

Climate issues are often approached as demanding technological innovation rather than better policy formation; however, this conservative treatment of knowledge creates a barrier to adaptation and threatens alternative ways of knowing. The magnitude of Arctic environmental change and its urgent consequences for society demand a disruption of the status quo. A more radical treatment of traditional ecological knowledge in policy documents could provide a counter-hegemonic discourse to the dominant construction of climate change (Pettenger, 2007, pp. 198–200).

Additionally, there are opportunities for scholars of the environment to dismantle the historically isolated practices of science in order to produce knowledge that is more readily usable in guiding adaptation. These approaches can work in tandem; already many scholars have called for greater inclusion of local knowledge in Arctic assessment reports. The challenge is then moving from knowledge to policy.

Analyzing the treatment of multiple knowledges in the AACA

Reports like the AACA are essential tools for the distribution of multiple knowledges because research conveyed through comprehensive assessments has more influence on policymaking (Brock, 2010, p. 133). It should be noted that the processes that resulted in the production of the document were not observed by this author, nor were contributors interviewed about their perspectives on the practices of collaborative research.

However, there is value in considering the text through this lens as that is precisely the information context in which policymakers operate. To this end, the following discussion will utilize the AACA summary for policymakers (titled *Barents Area Overview Report*) as the basis for textual analysis. The summary report condenses a 300-page report into 25 pages; obviously, a significant portion of the research is excluded. But what remains reveals the priorities and the agenda of the working group and a distillation of the ideas that are deemed critical.

Three questions are pursued in the discussion that follows:

1   How is knowledge in the AACA communicated in a way that is policy relevant?
2   How is traditional ecological knowledge treated in the AACA?
3   How does the AACA advance the project of integrating multiple knowledges into a coherent adaptation strategy?

As discussed, policy-relevant knowledge is conveyed in a way that is credible, legitimate and salient for the end users. In this case, the end users are identified as "the Arctic Council and the international science community" (AMAP, 2017a, p. vii). This is deliberately nonspecific in order to accommodate the wide range of possible users; one objective of the AACA is to introduce tools and strategies that can help inform decision-makers "in government, civil society, business and academia as they prepare to adapt to anticipated change in the Arctic" (AMAP, 2017b, p. 3). These end users may include local users, but the focus is on a larger scale of governance.

Consider, then, the characteristics that would make the AACA policy relevant and the context that would make it broadly usable. First, its credibility can be assessed by analyzing the validity and quality of the authors and sources of information. In the case of the AACA, credibility is derived from the diverse international list of authors. The report is structured to "promote stakeholder engagement, including participation from many different professional and public communities," which, if effective, corresponds to a high degree of legitimacy (AMAP, 2017b, p. 4). The emphasis on regional, decentralized expertise diminishes the likelihood that the report can be criticized for serving any particular interest group. Additionally, another element of credibility comes from the inclusion of Northern authors and the use of traditional knowledge, which provides a channel to local and indigenous end users.

The legitimacy of the report depends on the degree to which end users trust the institutions responsible for its production. The first sentence of the introduction of the summary for policymakers establishes this position; the report is clearly the product of the Arctic Council and thus relevant to the Arctic science and

decision-making communities (AMAP, 2017b, p. 4). In general, the Arctic Council is regarded as highly legitimate as the organization has two decades of success
in promoting international scientific assessments (Koivurova, 2010, p. 153). It has
a claim to legitimacy for indigenous users as well, given the practice of incorpo-
rating experts in TEK in the writing of the report. This ensures that there is at least
a basis for end users considering the AACA legitimate.

Saliency, which refers to the importance of information relative to the needs
of the end user, is perhaps the most difficult element to assess. Particularly when
considering issues related to adaptation, it is impossible to know what information users might need at any point in the future. In order to assess saliency,
we must consider "ecological, temporal, spatial, and administrative scales and
timelines . . . regulatory and legal constraints" as well as "values and beliefs of
stakeholders; the political landscape; and how information is communicated and
presented" (McNie, 2007, p. 20). The AACA targets many of these issues, and the
effort toward multidisciplinarity increases the likelihood that there is "something
for everyone." However, one could argue that the attempt to reach such a broad
array of end users makes the overall project *less* salient.

The researchers who produced the AACA were determined to generate
knowledge that was "useful and usable for making effective adaptation actions"
(AMAP, 2017b, p. 4). The report promises "key strategies and tools intended to
inform decision-makers about possibilities for helping their communities adapt
to future change" (AMAP, 2017a, p. ix). Whether or not the policy outcomes
match these goals, there is an explicit attempt to communicate in a way that is
policy relevant.

Given this stance, the content of the report could have significant implications.
The extent to which traditional ecological knowledge is authentically incorporated in the AACA has bearing on the long-term trajectory of adaptation knowledge. However, particularly in the summary for policymakers, there is a gap
between the stated value of TEK and the degree to which it is integrated with
science-based information. The first mention of indigenous peoples in the AACA
is not an explanation of how multiple knowledges are synthesized, nor is it a sug-
gestion of how local leaders can help their communities adapt to climate change;
it is a photograph of a nameless child from Yamal, Russia (AMAP, 2017b, p. 5).
The caption, "The tufts of reindeer fur on her hood are from the ears of the reindeer she owns," is a narrow treatment of indigeneity and offers little in the way of
substantive knowledge.

Likewise, the summary for policymakers has three sound bites that are set
aside from the primary text in quotations, seemingly to provide context and add a
human element to the report. Each of these text boxes addresses an issue relating
to the rights of indigenous peoples in the region. For example, Igor Slepushkin, a
Nenets reindeer herder from Yar-Sale, is quoted:

> Reindeer are the foundation of our life in the tundra. Thanks to our traditional
> knowledge accumulated over centuries while living in harmony with ani-
> mals, the land and the climate, we Nenets have kept our traditional lifestyle

of herding and thriving in the harsh climatic conditions of the Arctic, all the while our region is undergoing dramatic and in some cases, irreversible change.

(AMAP, 2017b, p. 12)

Pointing to this as an example in no way diminishes the challenges facing indigenous reindeer herders. The more critical point is the fact that this knowledge about past adaptation of the Nenets is textually separate. It is isolated, both conceptually and physically, from the main body of the report. The physical distinction of traditional ecological knowledge catalyzes an intellectual distinction as well; the reader sees the information as separate and, potentially, as less usable.

Both the visual elements and the text boxes demonstrate how traditional knowledge is included on a surface level, but its inclusion does not impact our understanding of knowledge writ large. Dominant Western knowledge is supplemented by other ways of knowing, but they remain textually outside of the mainstream. Even if multiple knowledges are integrated more authentically in the body of the report, this treatment effectively ensures that non-Western knowledge remains "other." This has a bearing on the sustainability of these forms of knowledge; while they will continue to be embedded in specific cultures, they may never be integrated into sustainable adaptation strategies.

The report does acknowledge that traditional knowledge has been underutilized in adaptation planning and proposes flexible strategies that can be used to strengthen the interface between science and policy (AMAP, 2017b, p. 20). The authors call for improvements in the way that indigenous knowledge is integrated in a broader science-policy interface; however, simply calling for the inclusion of local knowledge and advocating for moving debates about adaptation toward the public sphere do little to combat an otherwise conservative understanding of Arctic change.

Since the AACA was published less than two years ago, it is difficult if not impossible to determine the policy implications. However, it is possible to identify what the policy *goals* of the report are. The authors set out to lay foundations for future research, which they define as "key . . . elements that decision makers should consider in their work on adaptation" (AMAP, 2017b, p. 14). Six of the elements are informational; four offer suggestions for action. For each of these elements, I consider whether it provides a platform for integrating multiple knowledges and ask whether that platform has a high, moderate or low potential for advancing change (Table 8.1).

As shown, the adaptation foundations presented in the AACA summary for policymakers are highly dependent on the development and integration of new knowledge. In many cases, this knowledge is required to be multiple, meaning that it incorporates expertise from scientists, local community leaders, indigenous experts, national decision-makers and other social groups. This is not simply an intellectual exercise; it is essential for the development of effective adaptation strategies. Failing to reconfigure the construction of knowledge will all but ensure that adaptation in the North is reactive, rather than proactive or planned.

Table 8.1 The policy goals of the AACA vary in their potential to integrate multiple knowledges

Element	Potential	Explanation
Ecosystems and people in the area face a complex, interrelated range of impacts from climate and other drivers	High	This element suggests that there are numerous opportunities to engage with stakeholders with different backgrounds, skills and expertise. Diverse knowledges are critical for coordinating responses to change. Some of these responses will take place at the level of communities.
Adaptation actions must be placed in a broader context than climate change alone	Moderate	This element is clear about the need to situate adaptation as part of a broader sociopolitical process and articulates the imbalance of powers between local governments and communities as opposed to national governments and corporate actors.
Adaptation is a process, not an end in itself	High	This element emphasizes the need for cross-sectoral and holistic adaptation planning. It specifies that a high priority is to shift away from technical responses to specific climate change impacts and toward the social actors and institutions that enable holistic, cross-sectoral adaptation planning.
Resilience should be an objective of successful adaptation	Low	This element suggests that the ability to respond to a range of future impacts – resilience – is largely determined by individual communities. It discusses requirements for building resilience, but they are largely independent of the integration of multiple knowledges.
Integrating traditional and scientific knowledge is vital	High	This element calls for utilizing traditional and local knowledge in future planning steps and toward adaptation. They offer specific examples of how knowledge forms can be combined and what tools are available to identify key vulnerabilities and appropriate adaptive measures.
Tools exist to help decision-makers respond to uncertainty	Moderate	The tools referred to in this element are predominantly technical in nature and designed to facilitate near-term predictions about climate change. Examples are provided of models, scenarios, narratives and resilience indicators.
Understanding adaptation options requires an acknowledgment of barriers and limits	Moderate	This element articulates the need to understand the reasons behind barriers and limits to action. They include tradeoffs, insufficient guidance from national and regional authorities and the exclusion of traditional knowledge from local adaptation plans. They identify that existing knowledge about future climate change is inadequate but do not offer additional suggestions about how to deepen that knowledge base.

(*Continued*)

Table 8.1 (Continued)

Element	Potential	Explanation
Strengthened governance is necessary to manage adaptation processes	Low	This element reiterates the need for cross-sectoral and holistic adaptation planning and emphasizes its value in international, national and regional governance. It suggests that the Barents area has been a particularly effective arena for cooperation and advocates for "mainstreaming" of climate change adaptation into existing policy and governance.
It's necessary to strengthen the interface between science and policy	High	This element makes an explicit call for incorporating both scientific and traditional knowledge in order to generate holistic strategies. It suggests guidelines to support evidence-based local to Arctic-wide adaptation policy. Such guidelines would focus on usable knowledge, process-oriented approaches, long-term sustainability and communication across subcultures.
More knowledge is needed about current conditions and how they might change	Moderate	This element identifies specific knowledge gaps in how to develop and apply climate and socioeconomic scenarios. These gaps are important across nations, sectors and livelihoods. Usable knowledge about how climate change will interact with changing conditions is critical for the development of adaptation governance.

Conclusion

The AACA report illustrates a regional political ecology and provides a basis for understanding how multiple knowledges could be better integrated into large-scale assessments reports. The authors explicitly address the intersections of the climatic, demographic, economic, technology and political changes that will affect the Barents area. Moreover, they are clear that adaptive responses to these changes are contingent on an improved science-policy interface. Understanding the linkages between these issues requires researchers to engage with both local stakeholders and policymakers in a sustained and consequential manner. Pielke notes that this kind of work necessitates "the complex interface of science and decision making in which science is 'co-produced' by various sectors of society, and separation of 'facts' and 'values' cannot be achieved" (2004, p. 407).

This complex interface is the landscape in which northern researchers and policymakers are operating. Ultimately, adaptation decisions that are made without integrating multiple knowledges may lead to unsustainable results for those whose ways of knowing are outside the mainstream. It is far from obvious whose knowledge will lead to the most sustainable adaptation outcomes; therefore, the challenge is determining how to construct shared goals and strategies. The politics of adaptation are inextricably linked to the politics of knowledge. Both must transform in order to increase long-term sustainability for the people and environment of the Barents region and beyond.

References

AMAP (2014) *Adaptation actions for a changing Arctic: overall timeline.* Oslo: Arctic Monitoring and Assessment Programme (AMAP).

AMAP (2017a) *Adaptation actions for a changing Arctic: perspectives from the Barents area.* Oslo: Arctic Monitoring and Assessment Programme (AMAP).

AMAP (2017b) *Adaptation actions for a changing Arctic: Barents area overview report.* Oslo: Arctic Monitoring and Assessment Programme (AMAP).

AMAP (2019) *About AMAP* [Online]. Available at: https://www.amap.no/about (Accessed: March 1, 2019).

Arctic Council (2011a) *History of the Arctic Council* [Online]. Available at: https://arctic-council.org/index.php/en/about-us/arctic-council (Accessed: March 1, 2019).

Arctic Council (2011b) *Nuuk Declaration on the occasion of the Seventh Ministerial Meeting of the Arctic Council.* Nuuk, Greenland and Tromsø, Norway: Arctic Council.

Arctic Council (2018) *Index: observers* [Online]. Available at: https://arctic-council.org/index.php/en/about-us/arctic-council/observers (Accessed: March 1, 2019).

Barnhardt, R. and Kawagley, O.A. (2005) "Indigenous knowledge systems and Alaska native ways of knowing," *Anthropology & Education Quarterly*, 36(1), pp. 8–23.

Bourdieu, P. and Thompson, J. (1991) *Language and symbolic power.* Cambridge: Polity Press.

Brigham, L., Exner-Pirot, H., Heininen, L. and Plouffe, J. (2016) "Introduction—The Arctic Council: twenty years of policy shaping" in *Arctic Yearbook 2016.* Akureyri: Northern Research Forum, pp. 14–21.

Brock, D. (2010) "When is research relevant to policy making? A study of the Arctic Human Development Report," *Pimatisiwin: A Journal of Aboriginal and Indigenous Community Health*, 8(1), pp. 125–149.

Clark, W., van Kerkhoff, L., Lebel, L. and Gallopin, C. (2016) "Crafting usable knowledge for sustainable development," *PNAS*, 113(17), pp. 4570–4578.

Davis, A. and Wagner, J. (2003) "Who knows? On the importance of identifying 'experts' when researching local ecological knowledge," *Human Ecology*, 31(3), pp. 463–489.

Eriksen, S., Nightingale, A. and Eakin, H. (2015) "Reframing adaptation: the political nature of climate change adaptation," *Global Environmental Change*, 35, pp. 523–533.

Ford, J., Stephenson, E., Cunsolo Willox, A., Edge, V., Farahbakhsh, K., Furganl, C., Harper, S., Chatwood, S., Mauro, I., Pearce, T., Austin, S., Bunce, A., Bussalleu, A., Diaz, J., Finner, K., Gordon, A., Huet, C., Kitching, K., Lardeau, M-P., McDowell, G., McDonald, E., Nakoneczny, L. and Sherman, M. (2015) "Community-based adaptation research in the Canadian Arctic," *WIREs Climate Change*, 7, pp. 175–191.

Gorbačëv, M. (1987) *Speech in Murmansk at the ceremonial meeting on the occasion of the presentation of the Order of Lenin and the Gold Star to the City of Murmansk*, October 1, Murmansk, Russia.

Granderson, A. (2014) "Making sense of climate change risks and responses at the community level: a cultural-political lens," *Climate Risk Management*, 3, pp. 55–64.

Haas, P. (2004) "When does power listen to truth? A constructivist approach to the policy process," *Journal of European Public Policy*, 11(4), pp. 569–592.

Haas, P. (2016) *Epistemic communities, constructivism, and international environmental politics*. New York: Routledge.

IPCC (2014) "Annex II: glossary" in Mach, K.J., Planton, S. and von Stechow, C. (eds.) *Climate change 2014: synthesis report. Contribution of working groups I, II and III to the Fifth Assessment Report of the Intergovernmental Panel on Climate Change*. Geneva: IPCC.

Jasanoff, S. (eds.) (2004) *States of knowledge: the co-production of science and social order*. London: Routledge.

Koivurova, T. (2010) "Limits and possibilities of the Arctic Council in a rapidly changing scene of Arctic governance," *Polar Record*, 46(237), pp. 146–156.

Loring, P.A., Gerlach, S. and Penn, H.J. (2016) " 'Community work' in a climate of adaptation: responding to change in rural Alaska," *Human Ecology*, 44, p. 119.

McNie, E. (2007) "Reconciling the supply of scientific information with user demands: an analysis of the problem and review of the literature," *Environmental Science & Policy*, 10(1), pp. 17–38.

Moran, C., Russell, S.L. and Wishart, L.J. (2016) "Negotiating knowledge through boundary organisations in environmental policy" in Orr, K., Nutley, S., Russell, S.L., Bain, R., Hacking, B. and Moran, C. (eds.) *Knowledge and practice in business and organisations*. New York and London: Routledge, pp. 132–146.

Nadasdy, P. (1999) "The politics of TEK: power and the 'integration' of knowledge," *Arctic Anthropology*, 36(2), pp. 1–18.

NOAA (2018) *Arctic Report Card* [Online]. Available at: https://arctic.noaa.gov/ReportCard (Accessed: March 1, 2019).

Nyseth, T. and Viken, A. (2014) "Communities of practice in the management of an Arctic environment: monitoring knowledge as complementary to scientific knowledge and the precautionary principle?" *Polar Record*, 52(1), pp. 1–10.

Pettenger, M.E. (ed.) (2007) *The social construction of climate change*. Aldershot: Ashgate Publishing Limited.

Pielke, R. (2004) "When scientists politicize science: making sense of controversy over The Skeptical Environmentalist," *Environmental Science and Policy*, 7, pp. 405–417.

Royal Society (2010) *New frontiers in science diplomacy: navigating the changing balance of power*. London: The Royal Society.

Schlosberg, D., Collins, L. and Niemeyer, S. (2017) "Adaptation policy and community discourse: risk, vulnerability, and just transformation," *Environmental Politics*, 23(3), pp. 413–437.

Staudt, A., Leidner, A.K., Howard, J., Brauman, K., Dukes, J., Hansen, L., Paukert, C., Sabo, J. and Solórzano, L. (2013) "The added complications of climate change: understanding and managing biodiversity and ecosystems," *Frontiers in Ecology*, 11(9), pp. 494–501.

Young, O. (2002) *The institutional dimensions of environmental change: fit, interplay, and scale*. Cambridge, MA: MIT Press.

9 Arctic expertise and its social dimensions in Lapland

Monica Tennberg

Introduction

Finland is among the greenest, happiest, most innovative and most competitive countries in the world. Amid the international praise, Arctic expertise and smartness has become a popular term in national and regional discourses. The region-specific expertise builds on the human, social and creative capital of its peoples and their networks, natural resource-driven economic activities and their potential for economic growth and competitiveness in a global, knowledge-based economy. For Lapland, these discourses are full of promises in terms of international attention and resources for regional economic development. The questions thus are: How does the region do in assessments of human, social and creative capital? How does Arctic expertise contribute to knowledge economy in Lapland? An overview of different ways to assess social conditions, creativity and smartness in Lapland will follow a short introduction to the idea of Arctic expertise and its connections to social sustainability. The regional focus in this chapter is Lapland in domestic, Nordic, European and Arctic comparisons. These assessments use different sets of criteria and regional limitations but cover essential elements of Arctic expertise: decent living conditions and opportunities and critical mass of expertise as well as leadership, trust and networks.

Arctic expertise has multiple dimensions: first, it can refer to expert know-how, or procedural and often rather practical knowledge about how to live and work in the Arctic. The question is who knows *how*, but it is also a question of knowing *what*. Expertise can therefore also be understood as a declarative, more abstract knowledge of *knowing that*, pertaining to, for example, development in and adaptation to changing Arctic conditions. Expertise extends from rather automatic everyday practices, repetitive yet adaptive actions, to deliberate and planned practices of strategies and tactics. Expertise may be formally recognized with diplomas and certificates by universities or other educational institutions – although it is also possible for an expert to perform successfully and provide him/herself with a livelihood and recognition without any formal education. Expertise is also a collective resource, based on cooperation, networks and leadership with larger social and economic significance. Arctic expertise in Finland seems to include all the aspects of expertise and its various versions. Also, and importantly, it is claimed to be a national feature – Finnish expertise – situated in Lapland rather than a local facet such as Rovaniemi–based or regionally based expertise. However, various localities and regions claim their share of it (Prime Minister's Office, Finland, 2013, p. 15).

In the national discourse about Arctic expertise, concerns over social sustainability refer to adequate conditions for the health, working conditions and well-being of the population and improving resources to support it. In this approach, social sustainability should be included in impact assessments of major development projects and in research on social conditions in Northern Finland. Here, Arctic expertise builds on a competent, mobile and healthy work force but also on high levels of training and education and on investments in research and development as well as knowledge-based networks and partnerships. In order to develop this expertise further, investments are required for Arctic research and development and national and international collaboration (Prime Minister's Office, Finland, 2013, p. 20).

Similar ideas for the social basis of expertise in Lapland emerge from Lapland's program for Arctic specialization (Regional Council of Lapland, 2013). People in Lapland have adapted themselves and their livelihoods to the environment on the basis of "smart and arctic" knowledge, sustainable use of natural resources and strong communities. Smartness is envisioned as creating "high-level experts in Finland and in the EU on sustainable development of the northernmost regions" (Arctic Smartness Website, 2019). In this context, social sustainability refers to the maintenance of a strong competence base in production and service development: a healthy work force that is both competent and available. The smart specialization program highlights social sustainability and human, social and cultural capital as important aspects of Arctic expertise. This means that

> people, machines, equipment, companies and infrastructure, in fact the whole society, are capable of dealing with cold weather and sharp fluctuations in natural conditions. People, too, must be able to work efficiently and safely.
> (Regional Council of Lapland, 2013, p. 13)

The neighboring areas of Finnish Lapland similarly focus on social sustainability in their work on innovation and smart specialization. In Northern Sweden, regional expertise and smartness aims at an "Arctic Norrbotten–sustainable communities with world-class innovations" (*Norrbotten's political platform for Arctic specialization*, 2017, p. 13). The special nature of the region is recognized: "Together with northern Norway and northern Finland, northern Sweden and Norrbotten have the strongest population center in the Arctic region." Social sustainability also raises some concerns:

> These regions are facing demographic challenges with an ageing population, which has consequences for the labor market and public services. In general, people are living longer at the same time that young people are moving south. Attracting people to Arctic communities demands sustainable social planning and social innovations.
> (*Norrbotten's political platform for* Arctic *specialization*, 2017, p. 13)

Innovations include showing the potential of attractive residential and living environments; a culture of creativity and innovative thinking; opportunities for young people to realize their ideas and influence societal development to build an equal, open and multicultural Arctic society; increased access to services in both urban and rural areas and greater cross-border exchange in culture, sports and education (*Norrbotten's political platform for Arctic specialization*, 2017, p. 14). The social basis of expertise is also clear in Northern Norway:

> We will create economic growth and future-oriented jobs in the north in a way that takes account of environmental and social considerations. We will build local communities that can attract people of different ages and genders, and with different skills and expertise.
>
> (Norwegian Ministries, 2017, p. 3)

While Arctic expertise is generally perceived as a positive resource in Finland, there are tensions and controversies, too, over the social basis of Arctic expertise and smartness. The connections between Arctic expertise and social sustainability in Lapland will be investigated in this chapter from the perspective of governmentality (Foucault, 1991). The core concern of governmentality is populations and their well-being – in our case, social sustainability – and the different ways to define it and construct it as a part of Arctic expertise. Governmentality is a way to analyze how human, material and institutional relations are organized through practices of knowledge and power. Processes of identification, categorization and measurements of people and their characteristics are part and parcel of any effort in governmentality. The focus on regional expertise and smartness is a government-led exercise to motivate economic and other actors to pool their resources, align their interests and work together for economic growth, competition and development under the rather fuzzy umbrella of "Arctic expertise." As Mitchell Dean points out, governmentality is

> any more or less calculated and rational activity, undertaken by a multiplicity of authorities and agencies, employing a variety of techniques and forms of knowledge that seeks to shape conduct by working through our desires, aspirations, interests and beliefs, for definite but shifting ends and with a diverse set of relatively unpredictable consequences, effects and outcomes.
>
> (Dean, 1999, p. 11)

Social dimensions of expertise in Lapland

Lapland covers 30 percent of the Finnish land mass but has only about 3 percent (about 180,000 people) of the total population of about 5.4 million people. Traditional livelihoods and nature-based economic activities dominate the economy in Lapland. In the Finnish Arctic strategy (Prime Minister's Office, Finland, 2013, pp. 20–21), Arctic expertise divides the Northern population into two separate groups: "the Finnish Arctic population" and "the Sámi." The two groups are discussed separately in the strategy, with their own sets of governing aims and measures. This "biopolitical division" within the Northern populations is reflected in the development of indigenous specific indicators of social sustainability (Nymand Larsen et al., 2014) and community resilience (Carson et al., 2017). These indicators focus on sustainable livelihoods, nature contact and fate control, producing certain ways of governing people and their sustainability depending on the population they become categorized into (see also Hellberg and Knutsson, 2018). In the following, I will focus on the Finnish Arctic population.

Well-being of the population in Lapland

The Finnish Arctic strategy (Prime Minister's Office, Finland, 2013) makes a case for securing the living conditions for Northern populations. These conditions include both mental and material well-being, the opportunity to work, well-functioning basic services, equity, security and education. The basis for the population's well-being is economic development. Gross regional product (GRP) measures the overall economic output of all economic activities in a region. In the European context, the GRP per capita in Finnish Lapland (75–90) is lower than in the more Southern areas in Finland (90–100) and lower also than in the neighboring Nordic areas (100–125). The comparisons are based on the EU average for 28 countries (EU-28 = 100) (Eurostat, 2015).

The connection between economic development as indicated by GRP and social conditions is measured in different ways. The GINI co-efficient is a way to represent income equality or inequality in a country or an area, and is the most commonly used measurement of inequality. The lower the GINI co-efficient, the lesser the inequality in a country or an area. The average across OECD-EU countries is 0.30. The Nordic countries are among the most equal European countries. According to the World Bank (2018) data, the GINI co-efficient was 0.271 for Finland in 2012. For Lapland in 2012, it was 0.210–0.249 (Duhaime et al., 2017), which suggests that there is less inequality in Lapland than in the whole country. In addition to income equality, the United Nations Human Development Index (HDI) includes life expectancy and education in its assessment. The higher the index, the better the situation in a country. In 2018, Finland ranked 15th (0.920) in the global assessment of 189 countries, while Norway came 1st (0.953) and Sweden held the seventh place (0.933) (UN Nations Development Programme, 2018). The HDI for Lapland is lower than the national average due to lower levels of education, income and life expectancy.

The EU regional Social Progress Index (EU Commission, 2011) aims to measure social progress for each region as a complement to traditional measures of economic progress. The index is based on 50 indicators across three broad dimensions of social progress: basic human needs, foundations of well-being and opportunity. Here, Lapland does quite well (see Table 9.1). In a European comparison, Northern and Eastern Finland were ranked 11th among the 272 regions assessed in 2011. In terms of the foundations of well-being and opportunity, Lapland

Table 9.1 EU social progress index for Northern and Eastern Finland and Upper Norrland in Sweden, 2011

	Northern and Eastern Finland	Upper Norrland
Social progress index and rank in social progress	80.41/11th	82.33/1st
Rank in GDP per capita	118 (23,900 euros)	41 (31,500 euros)
Basic needs, score and ranking	82.69/105th	89.42/13th
Well-being, score and ranking	75.95/2nd	73.92/6th
Opportunity, score and ranking	82.70/13th	84.02/8th

Source: EU Commission (2011)

came 2nd and 13th, respectively, among the 272 sub-regions. It scored highly on access to information and communications, environmental quality, personal rights, personal freedom and choice and access to advanced education. It did less well on nutrition and basic health care, water and sanitation, shelter and personal safety. For basic human needs, Lapland ranked 105th. The neighboring region in Sweden, Upper Norrland, ranked 1st in the EU-wide assessment, scoring below average only in terms of access to medical care, traffic deaths and ICT (EU Commission, 2011).

These assessments send a mixed message. Income equality in Lapland is at the Nordic level, but in terms of social conditions, the general outlook is considerably lower in Finland than in the other Nordic and Arctic countries and even lower in Finnish Lapland.

Expertise and smartness as human resources

Knowledge-based economy in Lapland

In terms of a knowledge-based economy, most of Finland, according to the EU criteria, scores at an average level. The exception is the capital region in Helsinki and surrounding areas where the employment in high-technology and knowledge-intensive industries exceeds the EU average. According to this assessment, the Northern parts of Sweden and Norway also exceed the EU average (Nordregio, 2018).

According to the European regional innovation score board, RIS (EU Commission, 2018), Lapland is a "strong innovator." The Nordic average for innovation is 424 points, while Lapland gets 300. On the EU's scoreboard for regional innovation, the North and the East of Finland together are categorized as "strong innovators," Norrbotten and Västerbotten in Northern Sweden are among the top group as "innovation leaders" and Norwegian Finnmark, Troms and Nordland belong to the third level of "moderate innovators" (EU Commission, 2018).

In a more detailed and comparative analysis by Nordregio (Grunfelder et al., 2017, p. 51) on the Regional Potential Index (RPI) among the northern regions in Finland, Sweden and Norway, the scores vary between 110 in Lapland (Finland) and Murmansk (Russia) to 227.5 in Troms (Norway) (see Table 9.2). The RPI combines demographic, economic and labor force–based criteria. In this report, the region of Troms in Northern Norway stands out, especially on labor and economic dimensions. The lower RPI score for Lapland in comparison to Northern Norway and Sweden comes down to a relatively weak performance in both the labor and the economic dimensions of the RPI (Grunfelder et al., 2017).

The average score for Lapland on demographic indicators (45) is due to weak scores in population density, net migration and demographic dependency, but Lapland has a top score on female ratio: that is, more women in the region than in neighboring areas. Lapland's poor labor force score (30) is explained by the lowest employment rate and the highest youth unemployment rate. The employment rate in Lapland in 2015 was 66.1 percent, which is lower than the Nordic average (76.2 percent), the Finnish average (70.5 percent) and the EU average (67.1 percent). However, Lapland has a relatively educated labor force and a high share of the 25–64 age group with a higher education degree. For the economic dimension of the RPI, the average low score of Lapland (35) is due to relatively modest research and development investments and a relatively low value for gross regional product per capita. With R & D investments varying between 0.5 and 1.5 percent, Lapland does not meet the EU average of 2 percent investment in research and development. The smallest shares of knowledge-intensive jobs in Finland are located in Lapland and North and East Finland. In Lapland this is accounted for by the importance of basic and traditional industries (Grunfelder et al., 2017, pp. 51–54).

The role of entrepreneurship, companies and investments in research and development are considered critical in a knowledge-based economy. In Lapland, most companies are small and have limited opportunities for such investments. There are now 14,000 companies in Lapland, which translates to one company for 13 people in the region (variation between 13 and 24 people/company). The northern areas of Norway, Sweden and Finland have around 0.14 patent applications to the national office per 1,000 capita, while the rate is 0.22 for these

Table 9.2 Regional potential index by region, 2015

Ranking in 2015 (2010 ranking)	Region	Regional potential	Demographic criteria	Labor criteria	Economic criteria
1 (1)	Troms	227.5	67.5	80	80
2 (2)	Nordland	190	75	55	60
3 (3)	Norrbotten	172.5	37.5	50	85
4 (4)	Finnmark	135	45	55	35
5 (6)	Lapland	110	45	30	35
6 (5)	Murmansk	110	45	45	20

Source: Grunfelder et al. (2017, p. 51)

countries as a whole. Within the Northern regions, Northern Ostrobothnia in Finland has the highest average number of national patent applications at 81.2 per year, followed by the Swedish regions of Västerbotten with 47.5 and Norrbotten with 37.4. The other Northern regions have a much lower patent activity, ranging from 5.5 to 20 per year on average (Nordic Council of Ministers, 2018, pp. 20–23).

In order to make the most of Arctic expertise as economically significant and commercially viable products, services and innovations, the actors in the region are expected to mobilize their resources and to cooperate with each other. The notion of governmentality advocates governing at a distance. It promotes self-organization, which relies heavily on social capital and stresses the importance of cooperation and trust within institutions, companies and the state, as well as between individuals, for cooperation and development. The success of this approach calls for networks, partnerships and relations of trust. A recent study (Rinta-Kiikka et al., 2018) shows that the level of social capital in Finland is high but unevenly distributed. The highest social capital is found in South Ostrobothnia and Western (Varsinais-Suomi) and Central Finland (Kanta-Häme) while the lowest levels are in Lapland, Satakunta (in the Southwest) and Eastern Finland (North Karelia).

As a whole, social capital is lower in sparsely populated and remote areas (Rinta-Kiikka et al., 2018). This might be a critical issue in making Arctic expertise matter locally, regionally and nationally. Respect for and trust in one's own expertise and that of others was also required in a workshop in Rovaniemi, Lapland (Missä mennään Arctic Smartness yhteistyössä? November 6, 2018), when several expert and working groups came together to improve their strategic work and to develop different clusters. Ultimately, they organized themselves, their organizations and other bodies toward Arctic expertise. This suggests that, rather than the state directing others to do certain things, a complex set of apparatuses results in self-regulation among subjects and citizens to achieve the same ends (Rose and Miller, 1992, p. 184).

Creativity in Lapland

Finland's strategy for the Arctic region (Prime Minister's Office, Finland, 2013) and Lapland's smart specialization strategy (Regional Council of Lapland, 2013) both emphasize the Finnish Arctic human capital. Human capital, which is one of the capitals needed for economic development, refers to human resources and competencies, including skills, knowledge, education and vocational skills, leadership and creativity. The idea of human capital extends to intellectual capital, which includes access to and density of institutions such as universities, research institutions and schools.

Creative capital, according to Andrey Petrov (2014, p. 4), may be defined as "a stock of creative abilities and knowledge(s) that have economic value and are embodied in a group of individuals who either possess high levels of education and/or are engaged" in creative types of activities. Petrov identifies four central groups in the creative capital: entrepreneurs, leaders, scientists and artists. In his analysis, most Arctic regions have relatively weak creative capital, but many Arctic regions register remarkably high scores for leadership. Most Arctic regions exceed national baselines in the relative proportion of residents with occupations in arts and culture, which suggests a presence of cultural capital and a considerable potential of Arctic cultural economy, including traditional cultural economy. The circumpolar North has weak entrepreneurial capacities, especially in remote areas with large indigenous populations (Petrov, 2014, p. 6).

Petrov's analysis (2014) also contains data about populations with a university degree. On this "talent index," Lapland could do better: its performance is mediocre compared to other Arctic regions. In Lapland, the younger generations are clearly more educated than the older ones, and the level of education keeps rising. People with only basic education (nine years of comprehensive school) represent 17 percent of the 20- to-24-year-olds. There were clearly more people with secondary education in Lapland than the national average (+4.8 percent) and clearly fewer with university degrees than the national average (-4.9 percent) in 2014. In 2017, among those aged 15 and older, the highly educated made up 15 percent while in the whole country the number was 20.8 percent (Lapin liitto, 2014).

Petrov's leadership index refers to people in leadership and managerial occupations. There were 16,700 people in Lapland in 2017 employed as experts and holding leadership and managerial positions. In Finland as a whole, 490,000 people held leadership and managerial positions. Compared to the national average, more persons in Lapland are government or municipal employees, and more are in occupations related to teaching and education; 51 percent of them are women (53 percent in the whole of Finland) (Työvoimatutkimuksen palveluaineistot, 2018).

People with artistic and creative occupations – about 2,000 persons in Lapland or roughly 2.5 percent of all employed people in the region – are included in the so-called bohemian index. The whole of Finland has 105,801 people in these occupations, representing 4.24 percent of all employed persons. The share is clearly smaller in Lapland than the national average. However, the available statistics are quite old, dating back to the early 2010s (Lapin liitto, 2014).

Finally, the entrepreneurship index refers to people with business occupations, of whom there are 8,143 in Lapland (out of a total of 68,810 employed people in the region). While there were 12,120 companies in Lapland in 2015, the number has now risen to 14,000. Many of these are micro-companies with fewer than five employees. The total number of companies in Finland is 283,000. The workforce is employed mostly by private companies (49 percent), municipalities (30 percent), the state (8 percent) and state-owned companies (2 percent). Among them, there are also entrepreneurs (12 percent) (Lapin liitto, 2014).

Measured thusly, Lapland does not appear to have very significant creative capital. Compared to the Finnish national average, fewer people have university degrees, the number of artists is rather small, and most people work for the state, municipalities or state-owned companies. One can, of course, wonder what these numbers tell. The assessment highlights the number of people in certain positions and the importance of a critical mass for creativity. As Petrov's interest lies in the urban centers and their creativity, the administrative and educational capital of the county is Rovaniemi, with a population of 62,000. The city of Rovaniemi presents itself as a capital city of Arctic expertise, although by Petrovian standards, the creative capital in the city is rather modest. In the assessment of municipal competitiveness in 2016, Rovaniemi ranks only 52nd among the 73 assessed municipalities on criteria based on productivity, industrial development, entrepreneurship, education and innovation. Rovaniemi scores well only on the criteria of entrepreneurship and education (Satamittari, 2016). In 2012, the 3,378 Rovaniemi-based companies employed 13,000 people. The private sector is the largest employer (49 percent), and most of the companies are small. The economic structure is service-based. The municipality employs 27 percent of the labor force and the state 12 percent, while 9 percent are entrepreneurs (Rovaniemen kaupunki, 2018). There are two associations for professional artists, with approximately 200 members.

International mobility in Lapland

A key element in both the Finnish Arctic strategy (Prime Minister's Office, 2013) and Lapland's strategy for Arctic specialization (Regional Council of Lapland, 2013) is international mobility. The low fertility rates and aging populations are a national concern, not just an issue for Lapland only. The population of Lapland has kept decreasing for more than 20 years. In the midst of a dwindling and aging population and youth out-migration, international mobility is seen as a response and a part of developing Arctic expertise. International mobility is needed to compensate for the loss of population, workforce and taxpayers. In many ways, Lapland is already "an international region," with 4,000 foreign inhabitants (2017), 2.5 million tourists visiting the region yearly and hundreds of exchange students from Europe, Asia and North America. According to Finland's Arctic strategy (Prime Minister's Office, 2013, p. 13), "international networks, contacts and mobility are of great importance." The strategy notes (p. 21) that

> as it is challenging to find skilled labor in a sparsely populated area, the opportunities and problems associated with the mobility of the workforce will be felt more tangibly than further south.

Knowledge of local languages, Scandinavian languages, Russian and English "would offer employment opportunities for job-seekers across the entire Arctic region" (p. 21). In particular, the availability and free mobility of labor and businesses is seen as crucial for the economic growth and investments in Finland and in the Nordic countries.

Finland had about a quarter of a million foreign citizens in 2017. Since 2016, the numbers have increased nationally by 2.4 percent (5,813 people) but only by 0.5 percent (19 persons) in Lapland. Nationally, 4.5 percent of the population were foreign citizens in 2017, topped by Åland Islands (11 percent of the population) and followed by the capital area of Uusimaa, where 8 percent of the population were foreign citizens. In Lapland, foreign citizens made up 2.2 percent of the population. The share in Lapland of foreign citizens has grown in the last few years, almost doubling from 2005 to today. Kainuu and South Ostrobothnia have the lowest shares of foreign inhabitants, 1.9 and 1.6 percent, respectively. Among the cities, Helsinki leads with 61,000 foreign inhabitants, followed by Espoo and Vantaa with 29,000 and 25,000 foreign citizens. The city of Rovaniemi has 1,500 foreign citizens and the Kemi-Tornio area almost as many: 1,466 (2017). These two regions host most of the foreign citizens in the area, 3,000 out of 4,000. And while there are 241 refugees and asylum seekers in Lapland, the regional quota for refugees is larger (270). Most of the refugees and asylum seekers are located in the Kemi-Tornio area (77), followed by Rovaniemi (46) (Lapin ELY-keskus, 2018).

In recent years, the growth of the foreign population in Lapland has shown signs of slowing down. According to the immigration strategy of Lapland (Petäjä-maa, 2013), there should be 12,000 foreign citizens in Rovaniemi in 2030, which would require an annual growth rate of 7 percent. The current growth rate is only 0.5 percent, while in 2012–2014 it was 4 percent (Lapin ELY-keskus, 2018). The age structure of the current foreign population is younger than that of the general population. About a quarter of the foreign citizens in Lapland are unemployed, and 15 percent of them are considered long-term unemployed: that is, unemployed for more than a year. Their level of education varies: in 2017, 81 had basic education, 132 had secondary education and 79 had tertiary education. For 133 persons, the educational status was unknown (Lapin Ely-keskus, 2018).

In the neighboring regions across the national borders, the level of interna-tionalization is much higher. An average of 7.9 percent of the Nordic population are foreign nationals (Nordregio, 2017), ranging from 2 percent in Greenland to 11 percent in Norway. Nationally, Norway has a total of 799,797 foreign-born citizens while the corresponding figure in Sweden is almost 2 million (1,877,050 persons in 2017). The Swedish statistics use several population categories: for-eign citizens (897,336), foreign born with Swedish citizenship (1,064,041) and Swedish born with two foreign-born parents (561,957) (Statics Sweden, 2017). In Northern Norway, 29,222 foreign-born people lived in Troms and Finnmark in 2017 (Statistics Norway, 2018), and 16,889 foreigners had their residence in Norrbotten in Northern Sweden (Norrbotten county, 2017). Certain regions stand out as having a high share of foreign citizens in their populations, such as the Swedish part of the Torne River Valley bordering Finland.

But what kind of internationality and mobility is connected to the development of Arctic expertise? Despite the international character and the obvious appeal of the region for tourists, we need to ask what kind of internationalization would make visitors stay in Lapland and enhance the Arctic expertise, the knowledge-based economy and the competitiveness of the region.

Concluding remarks

The Finnish discourse on Arctic expertise takes its cue from the European Union discourse on regional smartness and innovations. Since the early 2010s, smart specialization has aimed to generate assets and capabilities based on the region's distinctive industrial structures and knowledge bases. The rationale for "smart specialization" is that by concentrating knowledge resources and linking them to a limited number of priority economic activities, countries and regions can become and remain competitive in the global economy. For the EU, this equals an Innovation Union, one of the initiatives in the Europe 2020 strategy for smart, sustainable and inclusive growth. The Innovation Union seeks to make Europe into a world-class science performer; remove obstacles to innovation, including skills shortages and revolutionize the way public and private sectors work together. The Innovation Union also includes social aims, such as 75 percent of people ages 20 through 64 to be working, 3 percent of the EU's GDP to be invested in research and development and at least 40 percent of people ages 30 through 34 having completed higher education. These efforts will help shape the EU as a part of the global "knowledge-based economy" in which knowledge in different forms, learning and innovation are central elements for societal and economic development (EU Commission, no date).

As a national and regional process, Arctic expertise is part of European "innovatization." National and regional efforts help make the idea of Arctic expertise reality. Collaborative processes at local, regional and national levels among authorities, companies and other stakeholders make Arctic expertise happen. The strategies of Arctic expertise and smartness also produce a new language for regional development with "clusters," "innovation and development environments" and supporting "ecosystems." In this discourse and practice of governing, the local, regional and national interests in Arctic expertise become nicely aligned in the face of diverse claims over its ownership and different approaches to utilizing it. Arctic expertise is locally based and regionally significant but "a common" national resource. In order to become naturalized, the internal tensions – such as ownership, values and expectations of cooperation and trust among partners – need to be settled. Through such processes of objectification, Arctic expertise is established as a resource to be developed and used within organizations, policies and plans. Naturalization makes Arctic expertise the normal order of things in everyday life without questioning the way it has been formulated and implemented (Duineveld et al., 2013).

While the importance of social sustainability as a part of Arctic expertise has been acknowledged in the key documents (Prime Minister's Office, Finland, 2013; Regional Council of Lapland, 2013), the later reports and debates have failed to cover social sustainability concerns and considerations for environmental sustainability. In the light of various criteria, indices and assessments for the social sustainability of Arctic expertise, the message is rather confusing. Depending on the selection of the criteria and regional focus (national, Nordic, European or Arctic), Lapland appears as "a strong regional innovator" or as "a county with opportunity" but also as "a county of weak creative and social capital" or as "an underperformer in knowledge-based economy." According to various social indicators produced by domestic, Nordic, European and Arctic assessments, Lapland's overall performance is poor compared to the neighboring areas in terms of basic needs, demography, economy and work force–related criteria. The social dimensions of Finnish Arctic expertise, at least for Lapland, do not appear very strong. The absence of social sustainability in regional and national debates in recent years suggests a particular Finnish governmentality of Arctic expertise. This engineering and bureaucratic approach calls for region-based expertise and innovations combined with capacity-building efforts but with little attention to the social aspects of expertise. But can there be expertise without experts? The Finnish model contrasts with the approaches adopted by the neighboring countries and their Northern areas working on their own understandings of Arctic expertise.

Also, indicators of social sustainability are not without problems of their own. There are issues of "calculative capture, problematic misalignments, and false measures of sustainability" as Thomson et al. (2014) point out. According to Yvonne Rydin (2007), indicators are a resource in national and international political games. An understanding of the relation between sustainability indicators and political practice of power is a key element to understanding today's sustainability indicators. The work to develop new criteria for social sustainability and indicators is a struggle over who receives the right to produce and direct the discourse and practice of sustainability. Indicators could ideally also be a political resource for local peoples and their communities or for regional decision-makers and stakeholders. The question is how local and regional actors construct themselves as subjects in response to indicator-based strategies of governing (Hellberg, 2017). The objectivity and scientificity of indicators are no longer core elements in explaining their success or failure. Instead, their success depends on their usefulness in terms of the exercise of power and knowledge. Indicators, often quantitative ones, establish a link between current popular discourses on knowledge and decision-making (see Rydin, 2007). They may challenge predetermined views, positions and solutions. In this case, my aim has been to raise awareness of concerns about social sustainability in the context of Arctic expertise.

References

Arctic Smartness Website (2018) Available at: https://arcticsmartness.eu/ (Accessed: January 30, 2019).

Carson, M., Sommerkorn, M., Larsen, R.K., Lawrence, R., Mustonen, T., Strambo, C., Vlasova, T. and Zhang, S. (2017) "A resilience approach to adaptation actions" in AMAP (Arctic Monitoring and Assessment Programme) (ed.) *Adaptation actions for a changing Arctic: perspectives from the Barents area.* Oslo: AMAP, pp. 195–218.

Dean, M. (1999) *Governmentality: power and rule in modern society.* London: Sage.

Duhaime, G., Caron, A., Lévesque, S., Lemelin, A., Mäenpää, I., Nigai, O. and Robichaud, V. (2017) "Social and economic inequalities in the Arctic" in Glomsrød, S., Duhaime, G. and Aslaksen, I. (eds.) *The economy of the North 2015.* Kongsvinger, Oslo: Statistics Norway, pp. 11–26.

Duineveld, M., Van Assche, K. and Beunen, R. (2013) "Making things irreversible: object stabilization in urban planning and design," *Geoforum*, 46, pp. 16–24.

EU Commission (2011) *EU social progress index 2011* [Online]. Available at: http://ec.europa.eu/regional_policy/en/information/maps/social_progress (Accessed: November 29, 2018).

EU Commission (2018) *EU regional innovation score board 2017* [Online]. Available at: http://ec.europa.eu/growth/industry/innovation/facts-figures/regional_en (Accessed: November 29, 2018).

EU Commission (no date). *Innovation Union* [Online]. Available at: https://ec.europa.eu/info/research-and-innovation/strategy/goals-research-and-innovation-policy/innovation-union_fi (Accessed: January 30, 2019).

Eurostat (2015) *Gross domestic product (GDP) per inhabitant in purchasing power standards (PPS) in relation to the EU-28 average, by NUTS 2 regions, 2015 (% of the EU-28 average, EU-28 = 100)* [Online]. Available at: https://ec.europa.eu/eurostat/statistics-explained/index.php?title=File:Gross_domestic_product_(GDP)_per_inhabitant_in_purchasing_power_standards_(PPS)_in_relation_to_the_EU-28_average,_by_NUTS_2_regions,_2015_(%25_of_the_EU-28_average,_EU-28_%3D_100)_MAP_RYB17.png&oldid=338468 (Accessed: January 30, 2019).

Foucault, M. (1991) "Governmentality," trans. Rosi Braidotti and revised by Colin Gordon, in Burchell, G., Gordon, C. and Miller, P. (eds.) *The Foucault effect: studies in governmentality*. Chicago: University of Chicago Press, pp. 87–104.

Grunfelder, J., Norlén, G., Mikkola, N., Rispling, L., Teräs, J. and Wang, S. (2017) *State of the Lapland region* [Online]. Stockholm: Nordregio. Available at: http://luotsi.lappi.fi/c/document_library/get_file?folderId=683161&name=DLFE-31404.pdf (Accessed: November 29, 2018).

Hellberg, S. (2017) "Water for survival, water for pleasure: a biopolitical perspective on the social sustainability of the basic water agenda," *Water Alternatives*, 10(1), pp. 65–80.

Hellberg, S. and Knutsson, B. (2018) "Sustaining the life-chance divide? Education for sustainable development and the global biopolitical regime," *Critical Studies in Education*, 59(1) [Online]. Available at: https://doi.org/10.1080/17508487.2016.1176064 (Accessed: April 8, 2019).

Lapin ELY-keskus (2018) *Ulkomaalaiset Lapissa* [Online]. Available at: https://view.officeapps.live.com/op/view.aspx?src=https%3A%2F%2Fely-keskus.fi%2Fdocuments%2F10191%2F15635979%2Flapely_Ulkomaalaiset_Lapissa.pptx%2F3ae0e99c-63b4-4fdb-b54c-6cb5c5c0930a (Accessed: November 29, 2018).

Lapin liitto [Regional Council of Lapland] (2014) *Lapin luotsi* [Lapin luotsi foresight] [Online]. Available at: http://luotsi.lappi.fi/1 (Accessed: November 29, 2018).

Nordic Council of Ministers (2018) *Arctic business analysis: creative and cultural industries* [Online]. Available at: http://norden.diva-portal.org/smash/get/diva2:1175681/FULLTEXT01.pdf (Accessed: November 29, 2018).

Nordregio (2017) *Foreign citizens as a share of total population in 2017* [Online]. Available at: www.nordregio.org/maps/foreign-citizens-as-a-share-of-total-population-in-2017/ (Accessed: January 29, 2019).

Nordregio (2018) *Employment in the high technology and knowledge intensive sectors 2016* [Online]. Available at: http://archive.nordregio.se/en/Maps/SNR18/Economy/Employment-in-the-high-technology-and-knowledge-intensive-sectors-2016/index.html (Accessed: January 30, 2019).

Norrbotten County (2017) *Population statistics, estimate 2017* [Online]. Available at: www.citypopulation.de/php/sweden-admin.php?adm1id=25 (Accessed: January 29, 2019).

Norrbotten's political platform for Arctic expertise (2017) [Online]. Norrbottens kommuner and Region Norrbotten. Available at: www.norrbottenskommuner.se/media/1956/norrbottens-political-platform-for-the-arctic.pdf (Accessed: November 29, 2018).

Norwegian Ministries (2017) *Norway's Arctic strategy: between geopolitics and social development* [Online]. Available at: www.regjeringen.no/contentassets/fad46f0404e14b2a9b551ca7359c1000/arctic-strategy.pdf (Accessed: November 29, 2018).

Nymand Larsen, J., Schweitzer, P. and Petrov, A. (eds.) (2014) *Arctic social indicators. ASI II: implementation.* TemaNord 2014:568. Copenhagen: Nordic Council of Ministers.

Petäjämaa, M. (2013) *Lapin maahanmuuttostrategia 2017* [Immigration strategy for Lapland 2017]. Elinvoimaa alueelle 1/2013. Rovaniemi: Lapin ELY-keskus.

Petrov, A. (2014) "Creative Arctic: towards measuring Arctic's creative capital" in Heininen, L., Exner-Pirot, H. and Plouffe, J. (eds.) *Arctic yearbook 2014: human capital in the north.* Akureyri: Northern Research Forum, pp. 149–166.

Prime Minister's Office, Finland (2013) *Finland's strategy for the Arctic region* [Online]. Government resolution on 23 August 2013. Prime Minister's Office Publications 16/2013. Available at: https://vnk.fi/documents/10616/1093242/J1613_Finland%E2%80%99s+Strategy+for+the+Arctic+Region.pdf/cf80d586-895a-4a32-8582-435f60400fd2?version=1.0 (Accessed: November 29, 2018).

Regional Council of Lapland (2013) *Lapland's Arctic specialisation programme. Lapland's smart specialisation strategy* [Online]. Available at: www.lappi.fi/lapinliitto/c/document_library/get_file?folderId=1483089&name=DLFE-21423.pdf (Accessed: November 29, 2018).

Rinta-Kiikka, S., Yrjölä, T. and Alho, E. (2018) *Talous, arvot ja alueellinen sosiaalinen pääoma.* PTT raportteja 258 [Online]. Available at: www.ptt.fi/media/julkaisut/rap258_full.pdf (Accessed: November 29, 2018).

Rose, N. and Miller, P. (1992) "Political power beyond the state: problematics of government," *The British Journal of Sociology*, 43, pp. 173–205.

Rovaniemen kaupunki (2018) Rovaniemen elinkeinotoiminnan raportointi vuodelta 2017. Konsernijaosto 13.2.2018. [Online]. Available at: http://rovaniemi.cloudnc.fi/download/noname/%7B372a515e-855a-4841-b5ae-0c0496fce442%7D/311109 (Accessed: August 21, 2019).

Rydin, Y. (2007) "Indicators as a governmental technology? The lessons of community-based sustainability indicator projects," *Environment and Planning D: Society and Space*, 25, pp. 610–624.

Satamittari (2016) *Suomen seutukuntien kilpailukyky* [Online]. Available at: www.satamittari.fi/suomen-seutukuntien-kilpailukyky-2016 (Accessed: November 29, 2018).

Statistics Norway (2018) *Foreign-born 1970–2018* [Online]. Available at: www.ssb.no/en/statbank/table/07109/ (Accessed: January 29, 2019).

Statistics Sweden (2017) *Summary of population statistics 1960–2017* [Online]. Available at: www.scb.se/en/finding-statistics/statistics-by-subject-area/population/population-composition/population-statistics/pong/tables-and-graphs/yearly-statistics – the-whole-country/summary-of-population-statistics/ (Accessed: January 29, 2019).

Thomson, I., Grubnic, S. and Georgakopoulos, G. (2014) "Exploring accounting–sustainability hybridisation in the UK public sector," *Accounting, Organizations and Society*, 39(6), pp. 453–476.

Työvoimatutkimuksen palveluaineistot, AKAVA (2018) Personal communication by email, October.

UN Nations Development Programme (2018) *Global human development indicators 2018* [Online]. Available at: www.indexmundi.com/facts/indicators/SI.POV.GINI/compare#country=fi (Accessed: November 29, 2018).

World Bank (2018) *Gini coefficient. Finland. World development indicators* [Online]. Available at: www.indexmundi.com/facts/indicators/SI.POV.GINI/compare#country=fi (Accessed: November 29, 2018).

10 When gender matters

Equality as a source of Arctic sustainability?

Heidi Sinevaara-Niskanen

Introduction

This chapter examines gender as an issue of resource and sustainability in the Arctic. In 2002, Mark Nuttall had already drawn attention to the question of resources and the ways in which resources were being perceived rather selectively within the Arctic. In his article, published in one of the long-standing polar journals *Polar Record*, Nuttall reflects Arctic capacities for sustainable development and the potential for building those capacities. Poignantly, he says that resources are more than (natural) resources that can be excavated, measured and consumed. According to Nuttall (2002, p. 194):

> sustainable development strategies often place emphasis on the 'resource' that is being used, exploited, mined or harvested, rather than on people and their social, cultural, and economic relationships with resources, the environment, and each other.

Indeed, in Artic science and politics, "resources" have been predominately treated as questions relating to the environmental impacts and economic significance of (the use of) natural resources (e.g. Glomsrød et al., 2017; Lempinen, 2017). Since the beginning of the 21st century, the widening scope of Arctic research has also created space for a more nuanced understanding of Arctic resources. Arctic research has morphed from a research field focused on environmental issues to a field of multidisciplinary studies interested in the well-being of the region and its people at large (Shadian and Tennberg, 2009). In a similar way, Arctic politics, which continues to be deeply intertwined with Arctic research, has awakened to observe the social and cultural developments taking place in the region. The Arctic human development reports, published under the auspices of the Arctic Council in 2004 and 2015, are concrete examples of the will of Arctic politics to know more about the region's social resources and (un)sustainabilities (AHDR, 2004, 2015).

The growing awareness of social "resources" has also paved the way for the discussion on the gendered nature of Arctic developments. Arctic science and politics alike have come to observe how patterns of Arctic demography, migration, education, well-being and, for example, questions of workforce are related to gender (e.g. AHDR, 2004, 2015). Arctic communities are tackling out-migration, lack of work opportunities, declining birth rates, challenges in accessing education and the growing need for social services, to mention just a few. The common denominator for these developments is, to a great extent, gender: the developments are *defined* by gender and *defining* the (future) social sustainability of the Arctic. Gender is a "prerequisite to sustainable development," as is also reminded by Nuttall (2002, p. 197).

Yet, while the significance of these developments has been acknowledged and there have been political (e.g. Gender equality in the Arctic, 2013; Taking Wing, 2002) and scientific (e.g. Signs, 2009; Vladimirova and Habeck, 2018) attempts to promote the topic, gender has remained a marginal theme in the discussions of Arctic development and sustainability. Questions of gender, as Hoogensen Gjørv (2017, p. 293) has observed when attending Arctic research conferences, continue to be present through their "distinct absence." In light of the continuing invisibility of gender, I ask: When have Arctic research and politics seen gender matter? By mapping the scope of social scientific Arctic research and the measures taken in Arctic politics – especially the work of the Arctic Council – this chapter offers a much-needed analysis of the ways in which questions of gender and sustainability mesh.

My focus will be on the discussions and developments that have taken place in the 21st century. In addition to providing an overview of the contemporary gender-specific Arctic research and policies, I will interrogate the lacks and silences entailed in them. My analysis builds on feminist theory and its discussions of equality. Although the particular focus is on gender, I will also touch upon other intersecting socio-cultural differences, such as age and ethnicity, as part of the engagement with feminist research. As my analysis reveals, even though questions of gender are not completely absent from Artic research and politics, they remain at the margins of the resource and sustainability considerations.

A brief Arctic history of adding gender and stirring

Feminist scholars have criticized the ways in which measures aimed to support gender equality are too often mechanical tactics of "adding women and stirring." Critical research has pointed out how including women in the prevailing social, political, legal and economic structures, without unpacking the gendered nature of those structures, provides very slim opportunities for gender equality to actually occur. For example, in political sciences, the debate around women's representation has gained wide attention, in particular the discussion of women's numeric representation and the impacts it might, could or is assumed to have on the substances of politics (e.g. Celis, 2008; Phillips, 1998; Squires, 2004). These discussions illustrate how, undoubtedly, representation matters. However, just adding women – or indigenous representatives, as has been the case in the structures of Arctic politics – does not automatically translate into politics being more sensitive to or inclusive of questions of gender or other socio-cultural differences, such as questions of ethnicity (Sinevaara-Niskanen, 2015a).

Indeed, providing room for alternative political and scientific agents and agendas requires more than just "adding and stirring." The title of this section should not, however, be read as a straightforward critique of Arctic politics and research not being responsive to issues of gender but as a note on how questions of gender have been latecomers in politics and research, in the Arctic as elsewhere. Within the history of environmentally driven Arctic research and politics, social scientific and human-centered approaches altogether have gained less attention.

The efforts to map, measure and protect the Arctic environment have had a decisive role in the establishment and the work of the Arctic Council. The council, as an intergovernmental decision-shaping body, has been and continues to be a significant producer and distributor of environmental knowledge of the Arctic (e.g. Hønneland and Stokke, 2007; Nilsson, 2012). Although the Arctic Council has created itself a particular "*cognitive* niche" through its engagement in science and knowledge (Stokke, 2007, p. 18), it took until the beginning of the 21st century for social developments to gain more attention. The Arctic human development report, published under the auspices of the Arctic Council in 2004, has been cited as one of the leading documents providing knowledge on the well-being and social and cultural life of Arctic residents. The report, as one of the first attempts to address social and cultural aspects of Arctic life, has laid the foundation for future knowledge endeavors on human development in the Arctic (Sinevaara-Niskanen, 2015b). In a similar way, when analyzing the history of global scientific collaboration efforts on polar research – that is, the International Polar Years (IPY) – Tennberg (2009) observed how the fourth IPY 2007–2008 finally marked a breakthrough for the integration of human dimensions into polar research.

For a long time, then, questions of gender, as other issues that had to do with human "life," were excluded from the agendas of polar research. In an article published in a special section on polar studies and gender, Rosner (2009, p. 490) illustrates how, even though gender was not on the agendas of Arctic research, Arctic research had a gender. As she vividly notes, "the grand heroic tradition of polar exploration" worked to define the regions "as all-male spaces of bonding, conquest, and noble suffering." Despite women's willingness to attend, female explorers were excluded from early polar voyages. Pointing out the paradox in this setting, Rosner (2009, p. 490) reminds that "women have lived in the Arctic for thousands of years." Yet for many years, and even for research, women's presence has gone "unnoticed."

What caught the attention of early social scientific Arctic research was the role of gender in the divisions of tasks and responsibilities in Arctic cultures and communities. Anthropologists in particular were interested in women's roles in traditional practices such as hunting and herding (Vladimirova and Habeck, 2018, p. 147). Vladimirova and Habeck (2018) note that although the scientific interest in gendered practices of the North had already emerged in the late 1970s, it took until the 21st century for Arctic feminist scholarship to appear. As they see it, feminist thinking has found its place within Arctic social scientific research only during the past decade.

The beginning of the 21st century has been equally significant in terms of Arctic politics and its engagement with discussions of gender and social sustainability. Two international conferences that have brought researchers and decision-makers together have been crucial. The first conference, Taking Wing, organized in 2002, has been referred to as the first Arctic event to discuss gender equality. The conference was arranged by the government of Finland during the Finnish chairmanship of the Arctic Council, and the aim was to specifically address questions of gender equality and women's lives in the Arctic (Taking Wing, 2002). What made the event particularly significant was that, for the first time, politicians, indigenous peoples, researchers and, for example, representatives of the member states of the Arctic Council gathered together to discuss the role of gender in Arctic developments. It is noteworthy that the main themes of the conference – women and work and violence against women, as well as gender and indigenous peoples' self-determination – have remained central in gender-specific Arctic research.

After the Taking Wing conference, it took more than ten years for the second gender-focused international Arctic conference to occur in 2014. The conference on Gender Equality in the Arctic (2014) was organized in Iceland, addressing topical issues such as resources, decision-making, political participation and well-being from a gender perspective. In comparison to the first conference, the 2014 event was a joint effort of several Arctic states. In addition, instead of being a single event within the current multitude of Arctic conferences, the conference was part of a larger project that seeks to raise awareness of gender equality. The project has an official status as an activity of the Arctic Council's Sustainable Development Working Group, which has as its ambitious task the realization of the Arctic Council's sustainable development goals, especially in terms of social and cultural dimensions. It could be argued that the 2014 conference and the project around it have succeeded in strengthening at least the institutional embeddedness of gender issues in Arctic politics.

Similarly, the Arctic human development reports (2004, 2015) have had a decisive role in facilitating discussions of gender and sustainability. These popularized scientific reports, born of science-policy interaction, have been perceived as highly influential in creating and upholding understandings of the social development of the Arctic. A feminist analysis of the 2004 Arctic human development report shows that the report depicts gender as something that has to do with women, a problem to be "solved" and an issue to be tackled. Instead of seeing gender as a positive resource, the report treats it as a source of vulnerability and a threat to social development (Sinevaara-Niskanen, 2015b). According to Hoogensen Gjørv (2017, p. 297), despite raising more questions than it provided analyses or answers, the first report succeeded in integrating "gender perspectives into Arctic research." However, her view on the second Arctic human development report, published in 2015, is more critical. The report settles for mainly stating that more gender-specific information is needed, even though gender is named as one of the report's crosscutting themes. Given the rather long time span in between the publication of the reports, Hoogensen Gjørv (2017, p. 298) concludes how the latest report gives very little evidence that "we have come much farther than the 2002 'Taking Wing' conference" (see also Svensson, 2017, for a public governance approach to the AHDR, 2015).

Analyzing the Arctic climate impact assessment – the leading report on Arctic climate change, published in 2004 – Martello (2008, p. 371) has also observed the partial (in)visibility of gender in the representations of indigenous peoples' climate knowledge. Her analysis points out that there is a male bias in both the understandings of indigenous peoples' climate observations and the visual representation of the peoples. Martello notes how the report tends to emphasize climate observations that are traditionally linked to male responsibilities, such as fishing and hunting, and how the photographs presenting indigenous persons in their surroundings depict mainly men.

Indeed, although the developments in Arctic research and politics seem encouraging, it is well worth examining more closely to what extent the political and academic discussions of the 21st century have included gender in the debates on Arctic sustainability. In order to dissect the ways in which gender intertwines with Arctic discussions of sustainability, I have focused on Arctic research and policy reports published between 2000 and 2018. The policy reports selected for the review are documents that are interested in questions of gender and have been produced under the auspices of the Arctic Council. In terms of mapping the ongoing research, I have concentrated on academic articles, books and book chapters written in English that have a specific focus on gender and that identify themselves as Arctic research. The requirement of an Arctic contribution, i.e. an outspoken engagement with the field of Arctic social sciences, has excluded many relevant nationally focused publications from the analysis (e.g. Keskitalo-Foley et al., 2013; Salmenniemi, 2008). In a similar way, a wide variety of academic research on gender and Northern issues written in languages other than English is not presented here (e.g. Autti et al., 2007). Therefore, although the studies collected for the analysis provide a good overview, they do not represent the whole range of research on gender in the Arctic.

Pushing towards equality: a feminist approach

My analysis of the data has been driven by a feminist approach. The development of feminist research and its arguments for recognizing questions of gender and achieving gender equality have been described through waves. These waves or phases of feminist thinking have emerged during different societal times and have had varying focuses in their calls for greater awareness and equality. The waves should not be considered as separate strands of discussion on the meaning of gender within society but rather as intersecting and complementary approaches to how to view questions of equality. The continuing coexistence of these waves also illustrates the necessity of having a multifaceted view on gender and equality – a question that is far greater than a simplified idea of "adding women."

The so-called *first wave*, led by suffragettes and feminist movements in the late 19th century, called for women's equal rights in politics and society at large. During the era when women had restricted access to many spheres of society, such as education, politics and legal institutions, the early feminists fought for women's rights to access education, to vote, to participate in politics and to be present in the public sphere. Thus, for first-wave feminism, questions of gender meant pushing toward equal societal access and opportunities for participation (Hesse-Biber, 2012; Matero, 1996). Although the majority of women today enjoy equal human rights, there is still a global need for the agendas of first-wave feminism. (Too) many girls and women of the world lack access to education, have limited opportunities for economic independence and struggle for equal political participation (Seager, 2008). Achieving gender equality continues to be a matter of social sustainability and development.

The *second wave* of feminism, often called the phase of standpoint feminism, emerged in the late 1970s. Supported by women's access to higher education and stronger engagement in science, feminist scholars started to question the male bias and masculine parameters embedded in knowledge and knowing. By critically unpacking whose knowledge our knowledge actually is and to what extent other types of knowledge are being excluded, second-wave feminists started to question the ways in which we "know." As a response to the practices of science and knowledge that were presented as universal yet had a male bias, second-wave feminism highlighted the need to understand knowledge always as contextual, specific, space-bound and corporeal. In particular, standpoint theories highlighted women's ways of knowing and argued for the need to recognize the subjectivity of knowledge (Harding, 1991; Hesse-Biber, 2012). The critique introduced by second-wave feminism continues to have a great impact, not only on practices of science, but also, more broadly, on how we perceive knowledge and knowing. Knowledge, and the acknowledgment of different types of knowledge, can work as a means of achieving equality. In light of the current global changes, politics has also recognized the significance of different types of, for example, environmental knowledge in the pursuit of sustainability. In particular, indigenous peoples' knowledge, which was historically excluded, has become celebrated as a route to a more sustainable world (Lindroth and Sinevaara-Niskanen, 2013).

The *third wave* of feminism has been driven by discussions of multiplicity. Born in the midst of the postmodern/post-structural turn of the late 1970s and early 1980s, the focus of third-wave feminism is on the study of a variety of socio-cultural categories, such as gender, race, age and religion, and their intersections. The black feminist movement in particular had a vital role in pointing out how feminist thinking had been dismissive of the significance of socio-cultural categories other than gender. Critical reflection on the relations of race and gender, and the ways in which racism is gendered (Smith, 1978), led to a broader discussion of "intersectionality" (Crenshaw, 1991). Introduced by feminists of color, the concept of intersectionality is linked to anti-racist and postcolonial struggles (Lykke, 2010, pp. 52–54; see also e.g. Weldon, 2008; Yuval-Davis, 2006). As Prah and Phoenix (2004, p. 76) define it, the approach focuses on the "multiple axis [sic] of differentiation – economic, political, cultural, psychic, subjective and experiential" and their intersections. It highlights how "different dimensions of social life cannot be separated out into discrete and pure strands" (Prah and Phoenix, 2004, p. 76). Through intersectionality, third-wave feminism has foregrounded the socio-cultural multiplicity of experiences and identifications that defines our positions within societies. The question of equality is thus more than a matter of gender; it is simultaneously a question of ethnicity, age and religion, among others. As St. Denis (2007) has also pointed out, when discussing indigenous women's perspectives on gender equality, ethnicity might well be considered as the most pertinent cause of inequality, not gender.

Arctic waves of equality

Arctic discussions of gender and equality entail all these waves of feminism. The "Arctic waves" that my analysis has revealed share similar features with the wider feminist debates, yet they raise specific place-bound concerns of equality and sustainability. Accordingly, the Arctic waves complement feminist development analyses that have tended to have their focus in the global South (e.g. Visvanathan et al., 2011). Indeed, if it has taken a while for Arctic research to address questions of gender, the same could be said about feminist research; it is only recently that Arctic people and phenomena have gained attention in feminist research.

The waves of gender-specific Arctic discussions presented here consist of calls for greater gender equality, arguments for the recognition of gender-specific vulnerabilities and an emphasis on (the need for) intersectional approaches. In particular, these waves illustrate the slow progress within Arctic research and politics when it comes to recognizing questions of gender. Little by little, gender has become "known" as a question of Arctic development and sustainability, although the pace and magnitude of that "knowing" has been much slower and slimmer than in global development discussions.

Having a say: on decision-making and resources

Debates concerning the management and development of the natural resources of the Arctic have been at the forefront of raising questions of gender in the Arctic. In 2004, the report on Arctic fisheries and women's participation in decision-making, published under the auspices of the Arctic Council, had already drawn attention to the declining role of women in natural resource management. The report noted how the global development of fisheries management and the fish processing industry had, in effect, privatized property that used to be common. As a result, women's role in the management of these resources had diminished, not only at the institutional level but also at the household and subsistence production levels (Sloan et al., 2004). Since the publication of that report, the discussion around resources and gender has been ever growing. Studies and policy reports have aimed at unpacking the gendered impacts of natural resource developments and figuring out the ways in which gender is, or should be, taken into account in decision-making processes (e.g. Mills et al., 2019; Skaptadóttir, 2004; Sloan et al., 2006; Staples and Natcher, 2015).

Scholarship in this vein has emphasized the need for greater equality, not only for the purpose of women having an equal share but also for the development of the region to be sustainable. Various case studies have pointed out how women have lacked access to the improvements and wealth linked to the development of natural resources in the Arctic. Women have not, for example, benefitted equally from the work opportunities or financial gains of these developments. There is also a great discrepancy in the management of natural resources; the management bodies are male dominated, and women lack equal possibilities for participation (e.g. Mills et al., 2019; Staples and Natcher, 2015). In particular, indigenous women's limited or unrecognized role in resource management has gained attention (e.g. Nightingale et al., 2017; O'Faircheallaigh, 2013; Stienstra, 2015). Even within traditional management bodies such as reindeer herding cooperatives, which are directly linked to indigenous cultures and livelihoods, indigenous women have faced challenges in participating in the decision-making (Kuokkanen, 2007).

The discussion of natural resources and gender has made visible how the lack of equality in resource governance is a crucial issue of sustainability. Gender equality has been demonstrated to strengthen sustainability in resource management, not only in its social sense but also in terms of its economic and environmental aspects. Limiting or denying women's access to natural resource management and to the opportunities that the use of those resources may provide can severely affect the development of the region at large.

Recognizing particular vulnerabilities

In addition to its natural resources, the Arctic region is commonly perceived through the threats it is (assumed) to face, be they environmental, social or economic. The scenarios of Arctic changes have tended to represent these risks as devoid of any particular socio-cultural connections, as if the (forthcoming) threats, and vulnerabilities that follow would be the same for everyone. Feminist discussions of Arctic vulnerabilities have, however, elaborated how Arctic vulnerabilities are, in effect, experienced differently. This line of scholarship highlights the gendered nature of, among others, issues of security, climate change and violence.

Feminist discussions of the impacts of climate change have brought to the fore the gender-specific impacts of environmental changes in the Arctic. For example, Naskali et al. (2016) have stressed the ways in which climate change affects especially elderly women. Eikjok (2007), from another point of view, has voiced the need to understand the social and cultural impacts that environmental changes have for indigenous men. Changes in traditional herding practices, or not being able to continue such practices, can greatly alter perceptions of masculinity in indigenous communities. In a similar way, feminist scholarship on Arctic security has emphasized the need to understand security in relation to individual experiences and interpretations. In particular, studies have pointed out how erasing violence and improving security in the Arctic requires recognition of gender-specific perspectives (e.g. Ingólfsdóttir, 2016; Irlbacher-Fox et al., 2014; Kuokkanen, 2015).

This strand of feminist Arctic studies reminds us that achieving sustainability requires recognition of gender-specific experiences and knowledge. As Hoogensen Gjørv (2017, p. 297) has summarized, "gendered social constructions and their resulting impacts on political, social and environmental life are not universal across Arctic geographies." Accordingly, as underlined by the discussions of gendered vulnerabilities, perceptions of Arctic threats and risks – often presented as universal – need to be replaced with more nuanced understandings of vulnerabilities. Knowledge is needed on the ways in which vulnerabilities are gendered.

Intersectional approaches

Reflections on how gender intersects with other socio-cultural categories emerged rather early on in feminist Arctic research. As also noted in the contexts of equality and vulnerability, gender becomes particularly significant when combined with other categories of identification. Even though one could describe the Arctic region through its multiple social, cultural, ethnic and religious differences, the struggle for gender equality in the midst of those differences has not been easy. In 2004, Eikjok described succinctly the challenges of equality and intersectionality in the Arctic:

> I have been involved in both the Sami ethno-political movement and the Sami women's movement since the 1970s. We were unpopular among our Sami brothers for introducing the women's cause into the struggle for our people's rights. We were unpopular among our fellow sisters in the wider community for bringing in our ethnic and cultural identity as women. Our brothers ridiculed us because gender was irrelevant for them; our Nordic sisters rebuked and lectured us because the minority and indigenous question was irrelevant for them. It has been difficult to fight for our rights both in relation to indigenous and feminist issues.
>
> (Eikjok, 2004, p. 57)

The past years have marked an increasing Arctic focus on intersectionality (e.g. Sinevaara-Niskanen, 2015a; Hoogensen Gjørv, 2017). A crosscutting approach on gender has also been applied in the science-policy reports published under the auspices of the Arctic Council (AHDR, 2015). Instead of being addressed as a separate issue, gender is now understood more in the context of other differences. The recent Arctic studies discussing multiple axes of difference have, for example, reflected on the relations of gender, religion and ethnicity (Valkonen and Wallenius-Korkalo, 2016), as well as on the connections of gender, ethnicity and sexuality (Olsén et al., 2018). In terms of violence, as elaborated by Kuokkanen (2015), the intertwinement of gender and ethnicity is highly significant. Indigenous women face violence in "disproportionate numbers," and "intersecting forms of racism and sexism combined with poverty and economic dependence" make them particularly vulnerable (Kuokkanen, 2015; p. 282). Indeed, as intersectional analyses have demonstrated, endeavors toward greater equality and sustainability in the Arctic should reach beyond "mere" matters of gender.

Conclusions

Given such developments in Arctic research and politics during the past 20 years or so, it would seem unfair to state that gender is (completely) lacking from Arctic agendas. Discussions of gender – the significance it has on Arctic developments and the role it plays, along with other socio-cultural differences, in the sustainability of the region – have found their place in both research and politics. It is noteworthy that this progress has not, and would not have, occurred without conscious efforts. Gender-specific Arctic research has had a significant role in providing much-needed knowledge. Following feminist discussions of gender equality, the Arctic waves have raised questions of equal participation, highlighted the pertinence of understanding gendered vulnerabilities and emphasized the importance of intersectional approaches.

Despite the progress and all the knowledge of the roles that gender plays in the region's development, questions of gender have remained at the margins of Arctic sustainability and resources debates. There is a growing awareness of the ways in which gender equality can support sustainability, how different types of gendered knowledge can assist in natural resource management, and, for example, why efforts to increase Arctic sustainability require responsiveness to a wide spectrum of socio-cultural aspects. Yet the position of natural resources in the discussions of Arctic resources remains unchallenged. Recognition of social resources, gender being one of them, still seems to lag behind. It is well worth problematizing this setting. The Arctic can be considered resource rich precisely because of its people and the social networks, cultural knowledge and place-bound practices that they have. As discussions of gender in the Arctic also point out, social sustainability contributes to the overall well-being of the region. In a similar way, gender should be seen as a source of wealth for the Arctic, a resource to draw on and an element of sustainability to be benefitted from.

References

AHDR (2004) *Arctic Human Development Report*. Akureyri: Stefansson Arctic Institute.
AHDR (2015) *Arctic Human Development Report. Regional processes and global linkages*. Copenhagen: Nordisk Ministerråd.
Autti, M., Keskitalo-Foley, S. Naskali, P. and Sinevaara-Niskanen, H. (eds.) (2007) *Kuulumisia. Feministisiä tulkintoja naisten toimijuuksista*. Rovaniemi: Lapland University Press.
Celis, K. (2008) "Gendering representation" in Goertz, G. and Mazur, A.G. (eds.) *Politics, gender, and concepts. Theory and methodology*. Cambridge: Cambridge University Press, pp. 71–93.
Crenshaw, K. (1991) "Mapping the margins: intersectionality, identity politics, and violence against women of color," *Stanford Law Review*, 43(6), pp. 1241–1299.

Eikjok, J. (2004) "Gender in Sápmi. Socio-cultural transformations and new challenges," *Indigenous Affairs* (1–2), pp. 52–57.

Eikjok, J. (2007) "Gender, essentialism and feminism in Samiland" in Green, J. (ed.) *Making space for indigenous feminism*. London: Zed Books, pp. 108–123.

Gender Equality in the Arctic (2013) Arctic Council, Sustainable Development Working Group, Project [Online]. Available at: https://arcticgenderequality.network/ (Accessed: March 25, 2019).

Gender Equality in the Arctic (2014) *Current realities, future challenges*. Conference Report [Online]. Iceland: Ministry for Foreign Affairs. Available at: www.stjornarradid.is/media/utanrikisraduneyti-media/media/nordurslodir/gender-equality-in-the-arctic.pdf (Accessed: March 25, 2019).

Glomsrød, S., Duhaime, G. and Aslaksen, I. (eds.) (2017) *The economy of the North 2015*. Oslo: Statistics Norway.

Harding, S. (1991) *Whose science? Whose knowledge? Thinking from women's lives*. Ithaca: Cornell University Press.

Hesse-Biber, S.N. (2012) "Feminist research: exploring, interrogating and transforming the interconnections of epistemology, methodology and method" in Hesse-Biber, S.N. (ed.) *Handbook of feminist research*. London: Sage Publications, pp. 2–26.

Hønneland, G. and Stokke, O.S. (2007) "Introduction" in Hønneland, G. and Stokke, O.S. (eds.) *International cooperation and Arctic governance. Regime effectiveness and northern region building*. New York: Routledge, pp. 1–12.

Hoogensen Gjørv, G. (2017) "Finding gender in the Arctic: a call to intersectionality and diverse methods" in Latola, K. and Savela, H. (eds.) *The interconnected Arctic—UArctic congress 2016*. Cham: Springer, pp. 293–303.

Ingólfsdóttir, A.H. (2016) *Climate change and security in the Arctic. A feminist analysis of values and norms shaping climate policy in Iceland*. Rovaniemi: Lapland University Press.

Irlbacher-Fox, S., Price, J. and Wilson Rowe, E. (2014) "Women's participation in decision-making: human security in the Canadian Arctic" in Hoogensen Gjørv, G., Bazely, D.R., Goloviznina, M. and Tanentzap, A.J. (eds.) *Environmental and human security in the Arctic*. London: Routledge, pp. 203–230.

Keskitalo-Foley, S., Naskali, P. and Rantala, P. (eds.) (2013) *Northern insights. Feminist inquiries into politics of place, knowledge and agency*. Rovaniemi: Lapland University Press.

Kuokkanen, R. (2007) "Myths and realities of Sami women: a post-colonial feminist analysis for the decolonization and transformation of Sami society" in Green, J. (ed.) *Making space for indigenous feminism*. London: Zed Books, pp. 72–107.

Kuokkanen, R. (2015) "Gendered violence and politics in indigenous communities," *International Feminist Journal of Politics*, 17(2), pp. 271–288.

Lempinen, H. (2017) *The elusive social: remapping the soci(et)al in the Arctic energyscape*. Rovaniemi: Lapland University Press.

Lindroth, M. and Sinevaara-Niskanen, H. (2013) "At the crossroads of autonomy and essentialism: indigenous peoples in international environmental politics," *International Political Sociology*, 7(3), pp. 275–293.

Lykke, N. (2010) *Feminist studies. A guide to intersectional theory, methodology and writing*. New York: Routledge.

Martello, M.L. (2008) "Arctic indigenous peoples as representations and representatives of climate change," *Social Studies of Science*, 38(3), pp. 351–376.

Matero, J. (1996) "Tieto [Knowledge]" in Koivunen, A. and Liljeström, M. (eds.) *Avainsanat. 10 askelta feministiseen tutkimukseen*. Tampere: Vastapaino, pp. 245–269.

Mills, S., Dowsley, M. and Cox, D. (2019) "Gender in research on northern resource development" in Southcott, C., Abele, F., Natcher, D. and Parlee, B. (eds.) *Resources and sustainable development in the Arctic*. London: Routledge, pp. 251–270.

Naskali, P., Seppänen, M. and Begum, S. (eds.) (2016) *Ageing, wellbeing and climate change in the Arctic: An interdisciplinary analysis*. London: Routledge.

Nightingale, E., Czyzewski, K., Tester, F. and Aaruaq, N. (2017) "The effects of resource extraction on Inuit women and their families: evidence from Canada," *Gender & Development*, 25(3), pp. 367–385.

Nilsson, A.E. (2012) "Knowing the Arctic: the Arctic Council as a cognitive forerunner" in Axworthy, T.S., Koivurova, T. and Hasanat, W. (eds.) *The Arctic Council: its place in the future of Arctic governance*. Toronto: Walter & Duncan Gordon Foundation, pp. 83–112.

Nuttall, M. (2002) "Global interdependence and Arctic voices: capacity-building for sustainable livelihoods," *Polar Record*, 38(206), pp. 194–202.

O'Faircheallaigh, C. (2013) "Women's absence, women's power: indigenous women and negotiations with mining companies in Australia and Canada," *Ethnic and Racial Studies*, 36(11), pp. 1789–1807.

Olsén, L., Heinämäki, L. and Harkoma, A. (2018) *Human rights and multiple discrimination of minorities within minorities: Sámi persons with disabilities and sexual and gender minorities*. Juridica Lapponica 44. Rovaniemi: University of Lapland.

Phillips, A. (1998) "Democracy and representation: or, why should it mater who our representatives are?" in Phillips, A. (ed.) *Feminism & politics*. Oxford: Oxford University Press, pp. 224–240.

Prah, A. and Phoenix, A. (2004) "Ain't I a woman? Revisiting intersectionality," *Journal of International Women's Studies*, 5(3), pp. 75–86.

Rosner, V. (2009) "Gender and polar studies: mapping the terrain," *Signs: Journal of Women in Culture and Society*, 34(3), pp. 489–494.

Salmenniemi, S. (2008) *Democratization and gender in contemporary Russia*. New York: Routledge.

Seager, J. (2008) *The Penguin atlas of women in the world*. London: Penguin Books.

Shadian, J.M. and Tennberg, M. (2009) "Introduction" in Shadian, J.M. and Tennberg, M. (eds.) *Legacies and change in polar science. Historical, legal and political reflections on the International Polar Year*. Surrey: Ashgate, pp. 1–8.

Signs (2009) "Comparative perspectives symposium: gender and polar studies," *Signs: Journal of Women in Culture and Society*, 34(3), pp. 489–532.

Sinevaara-Niskanen, H. (2015a) *Setting the stage for Arctic development: politics of knowledge and the power of presence*. Rovaniemi: Lapland University Press.

Sinevaara-Niskanen, H. (2015b) "Vocabularies for human development: Arctic politics and the power of knowledge," *Polar Record*, 51(2), pp. 191–200.

Skaptadóttir, U.D. (2004) "Responses to global transformations: gender and ethnicity in resource-based localities in Iceland," *Polar Record*, 40(3), pp. 261–267.

Sloan, L., Kafarowski, J., Heilmann, A., Karlsdóttir, A., Aasjord, B., Udén, M., Öhman, M-B., Singh, N. and Ojalammi, S. (2006) *Women and natural resource management in the rural north*. Norfold: Kvinneuniversitetet Nord.

Sloan, L., Kafarowski, J., Heilmann, A., Karlsdóttir, A., Udén, M., Angell, E. and Moen Erlandsen, M. (2004) *Women's participation in decisions-making processes in Arctic fisheries resource management*. Norfold: Kvinneuniversitetet Nord.

Smith, B. (1978) "Toward a black feminist criticism," *The Radical Teacher*, (7), pp. 20–27.

Squires, J. (2004) *Gender in political theory*. Cambridge: Polity Press.

St. Denis, V. (2007) "Feminism is for everybody: aboriginal women, feminism and diversity" in Green, J. (ed.) *Making space for indigenous feminism*. London: Zed Books, pp. 33–52.

Staples, K. and Natcher, D.C. (2015) "Gender, decision making, and natural resource co-management in Yukon," *Arctic*, 68(3), pp. 356–366.

Stienstra, D. (2015) "Northern crises," *International Feminist Journal of Politics*, 17(4), pp. 630–651.

Stokke, O.S. (2007) "Examining the consequences of Arctic institutions" in Stokke, O.S. and Hønneland, G. (eds.) *International cooperation and Arctic governance. Regime effectiveness and northern region building*. New York: Routledge, pp. 12–26.

Svensson, E-M. (2017) "Feminist and environmentalist public governance in the Arctic" in Körber, L-A., MacKenzie, S. and Westerståhl, A. (eds.) *Arctic environmental modernities. From the age of polar exploration to the era of the Anthropocene*. Cham: Palgrave Macmillan, pp. 215–230.

Taking Wing (2002) *Conference Report*. Helsinki: Ministry of Social Affairs and Health.

Tennberg, M. (2009) "Three spirals of power/knowledge: scientific laboratories, environmental panopticons and emerging biopolitics" in Shadian, J.M. and Tennberg, M. (eds.) *Legacies and change in polar science. Historical, legal and political reflections on the International Polar Year*. Surrey: Ashgate, pp. 189–200.

Valkonen, S. and Wallenius-Korkalo, S. (2016) "Practising postcolonial intersectionality: gender, religion and indigeneity in Sámi social work," *International Social Work*, 59(5), pp. 614–626.

Visvanathan, N., Duggan, L., Wiegersma, N. and Nisonoff, L. (eds.) (2011) *The women, gender and development reader*. London: Zed Books.

Vladimirova, V. and Habeck, J.O. (2018) "Introduction: feminist approaches and the study of gender in Arctic social sciences," *Polar Geography*, 41(3), pp. 145–163.

Weldon, S.L. (2008) "The concept of intersectionality" in Goertz, G. and Mazur, A. (eds.) *Politics, gender and concepts: theory and methodology*. Cambridge: Cambridge University Press, pp. 193–218.

Yuval-Davis, N. (2006) "Intersectionality and feminist politics," *European Journal of Women's Studies*, 13(3), pp. 193–209.

11 Sámi cultural heritage and tourism in Finland

Francis Joy

Introduction

This chapter discusses the use of indigenous traditions from the material culture and illustrates how the copying and manufacturing of art taken from revered divination drums and their symbolism (cosmological landscapes) appropriate the culture of the Sámi people of Fennoscandia and the Kola Peninsula. The drums are sacred cultural artifacts that have their origins in a pre-Christian religion that encountered Christianity predominantly in the 14th century when Finland was a part of the kingdom of Sweden. Artifacts as such and their symbolism are appropriated into tourism for many reasons. The main ones can be attributed to an overall lack of protection, a loss of culture through colonialism, ethno-tourism and lack of education concerning the cultural heritage of the Sámi people in the education systems in general throughout Fennoscandia and the Kola Peninsula.

Around 70 drums that were taken away from Sámi *Noaiddit* ("shamans") by missionaries and priests in the 17th century have survived, and hundreds more were confiscated and burned. The looted drums have since been in the custody of museums in France, Germany, Britain, Italy, Denmark and Sweden, to which the Sámi have little or no access. While there are ongoing repatriation issues, in most cases the Sámi have to rely on photographs of these artifacts that are an essential part of their cultural heritage.

The symbolism and figures of the drums are considered to be sources of traditional knowledge for the Sámi people and representative of the collective cultural memory and heritage. When understood from an economic perspective, they are sustainable sources of income and employment, and their reproduction helps preserve Sámi culture and traditions. Perhaps this is best described by Sunna (2006, pp. 24–25), who explains how these practices are a natural basis for Sámi culture:

> There are older drums from the 17th and 18th centuries and newly made ones. The drum is an important link to the past. It preserves ideas of our concepts of the world, of our protectors, and the religion practiced before the encroachment of the church. The artist or artisan uses the older holy symbols as a link to the past. The symbols can also serve as a reminder of that which once happened and inspire a trip from past to present. Sámi mythology and religion are

an invaluable source for the Sámi artist and artisan. During the last half of the 20th century many have exhibited brilliant work that has taken its inspiration from the Sámi religion.

In some cases, these artifacts and their symbolism have been copied and recreated by Sámi artists for the purposes of relearning about their own culture and history. The drums enable reconnection with the history of Sámi ancestors and transmit these aspects of Sámi culture across families and generations to maintain the health and well-being of the community. One further reason for recreating drums and reusing ancient symbolism is to educate outsiders about Sámi culture. Typically, this forwarding of traditional knowledge contributes to living a good life; to building, maintaining and sustaining identity and to providing a sense of security. Cultural heritage and its reuse among the Sámi can be seen as a reflection of the values within nature and the vitality of the culture where cosmological structures painted on drums reflect the spiritual relationship with the land. They convey principles and virtues that represent how established relationships with animals and supernatural powers have existed for a very long time.

For the Sámi people, when traditional knowledge is created as handicraft products, the products are called *Duodji* and are protected by the *Duodji* trademark, which is owned by the Sámi Council.

> By *Duodji* we mean the handicrafts and artistic handicrafts made by the Sámi based on Sámi traditions. Sámi design, Sámi patterns and colours. The word *Duodji* is also used as a mark of authentic Sámi handicraft.
>
> (Sunna, 2006, p. 5)

In the creation and use of a drum, there is a range of customs and taboos to be adhered to. These not only apply to the making of the sacred drums, but also determine the decoration and usage of natural products, such as bone, antlers and reindeer hide. Violation of these principles is considered to bring about bad luck and adversity.

By contrast, and throughout Finland, fake artifacts that are deemed to be tied to Sámi cultural heritage and manufactured both within local industries and among Sámi artists are used as a touristic marketing tool to make money, thus highlighting the adaptation and subsequent exploitation of Sámi traditions. The reuse and retailing of Sámi traditions take place throughout Fennoscandia and the Kola Peninsula but is much more of a problem in Finland because the production is industrial in its scale and is subsequently sold in hotels, department stores and souvenir shops all over the world.

The use and symbolism of drums can be linked with the study of Sámi religion as tourism companies offer exotic shamanistic encounters and even pilgrimage-like wilderness journeys. By claiming to provide something genuine and original, these companies make use of Sámi cosmology from drums, *joiking* (a special form of Sámi singing) and visits to sacred sites as virtual learning experiences that offer a break from everyday mundane reality.

The study of this subject matter is important for several reasons. First, it brings into focus a number of voices raised within the Sámi community, addressing what Sámi people feel amounts to inappropriate and harmful use of Sámi cultural heritage in tourism. Second, these Sámi voices draw attention to a general lack of protection and the failure of both national and international legislation and implementation of guidelines for the tourism business to follow. Third, it is conceivable that one of the main driving forces behind appropriation is the belief that Sámi pre-Christian religion no longer exists and that Sámi traditions are therefore obsolete. This makes appropriation freely available.

Broader discussions about appropriation of Sámi cultural heritage within Sámi culture

My research can be linked with similar themes addressed in other scholarly works on appropriation of Sámi cultural history, landscapes, art and handicrafts. These works have drawn attention to both the implications and the consequences for the Sámi and their heritage and have helped broaden discussions of cultural appropriation and sustainability. In particular, this pertains to Sámi religion and shamanism, as well as the function and representation of sacred *noaidi* drums and the tradition of *Duodji* (Mathisen, 2010; Bydler, 2017; Guttorm, 2007).

In exploring indigenous spirituality in the touristic border zone, Mathisen (2010) focuses on various Sámi cultural landscapes and historical contexts, and examines

> how a specific version of Indigenous spirituality and religion is presented in a contemporary touristic display, taking the *Sápmi* theme park's exhibition of pre-Christian Sámi religious myths and spirituality as a point of departure in the village of *Kárášjohka*, in north-eastern Norway.
>
> (Mathisen, 2010, p. 54)

Mathisen makes relevant points about problematic issues surrounding ownership, the utilization of cultural heritage within tourism and its consequences regarding exploitation of Sámi culture. All this "influences contemporary understandings of indigeneity and spirituality" (Mathisen, 2010, p. 59). He likewise discusses images of drums that are used as "sign posts . . . emphasizing the route visitors are about to follow" (Mathisen, 2010, p. 54) in the theme park. The study also raises a series of questions regarding the influence of Western culture and mentality on the customs of the indigenous Sámi in the town that further colonize rather than decolonizing spiritual traditions.

To further highlight problems within Sámi culture regarding the reuse of traditional knowledge, de Jong (2017) has studied Sámi identity representation and revitalization in northern Norway. She found that

> when designing this park, the Sámi parliament approved the plan of using Sámi drums, a sacred symbol, as signs for the toilets. They hang the drums up with "Toilet" written on it.
>
> (De Jong, 2017, p. 76)

The Swedish scholar Charlotte Bydler (2017) has similarly examined various complexities involved in cultural property and ownership in relation to the *Duodji* handicraft tradition and its protection. She also seeks to identify the roots of appropriation:

> [D]espite constant speculation that Duodji must be protected lest it become extinct, this has failed to happen. Instead, *Duodji* has been institutionalized and professionalized for almost six decades now. Arguably, an important impulse came from the nineteenth-century sloyd revival that was part of the European majority-societies' vogue for national romanticism. The arts & crafts movement, a counter-modernist force in Britain, inspired zealous majority-Sweden (and likeminded people in the rest of the white world) to organize crafts in regional centers that mapped material and immaterial culture onto practitioners and regions – except, according to Lilli Zickerman, for the Sámi *Duodji*, which was considered beyond the pale of development and only good for tourists (Lilli Zickerman, Hemslöjdkommitténs betänkande, 1918, p. 230) quoted in Hyltén-Cavallius, 2014.
>
> (Bydler, 2017, p. 153)

Bydler's contribution is important for the study because from an assessment of the sources I have located concerning the appropriation of drums and their symbolism into tourism, it has been difficult to ascertain when and where this began. However, and although a speculative assessment, it seems that somewhere between the 1990s and the early part of the 2000s would be a rough estimation of when the development of tourism increased.

Gunvor Guttorm, a Norwegian Sámi professor of *Duodji*, is an authority within the Sámi community on the production, use and application of *Duodji*. In 2007, she published a chapter in a book titled "*Duodji* – Sami Handicrafts – Who Owns the Knowledge and the Works?" Guttorm makes a number of important points about the meanings, reuse and ownership of drum symbols and their application by Sámi craftspeople in new artistic creations. She also addresses issues concerning cultural property and the retailing of false *Duodji* by Sámi craftspeople in Norway as well as the "adaptation of Sámi symbols within tourism in Finland" (Guttorm, 2007, p. 84). There are clear and explicit guidelines for the production and sale of *Duodji*, which some Sámi adhere to while others do not. What is certain is that the subject matter is highly complicated and implies the devaluing and corruption of Sámi traditional knowledge and traditions in some contexts while elevating them in others.

Other contributing factors which play a role in appropriation

In every gift shop throughout *Sápmi*, one can find a broad range of souvenirs that are fake replicas of Sámi *Duodji*. These range from copies of sacred divination drums to Sámi traditional costume (*Gákti*) to a whole range of household goods, clothes, textiles and jewelry that exhibit religious symbolism copied from drums which were and are still used by *Noaiddit* for their sacred function.

While small businesses in Finland have to contend with high sales taxes and struggle to make a living, their unsustainable cultural practices of appropriating Sámi cultural heritage exploit Sámi culture and traditional pre-Christian religion in various dysfunctional ways. They can therefore be seen as harmful because they destroy the underlying values. The problem is the concept of freedom of trade and marketing, which turns Sámi traditions into tourism and hurts the Sámi people and their society as a whole. Obviously, such practices are not only widespread in Finland and the small businesses run by Sámi artists – causing internal conflicts among the Sámi – but Sámi heritage is also manufactured in China and Estonia and then shipped back to Finland and other Nordic countries. This gives some idea of the complexity of the subject matter at hand.

As a way of further demonstrating the complexities involved in the appropriation of Sámi culture, it is important to take into consideration some of the other mechanisms that can be identified as contributory factors regarding the marketing of Sámi spiritual culture through tourism in Finland. A good place to start is the historical background and outlawing of Sámi religion predominantly since the 14th century when Finland was a part of Sweden and the 16th century when *Noaiddit* were put to death in Norway and Sweden for alleged sorcery pertaining to the use of the drum. As a result of over 600 years of colonialism, a general belief prevails outside of Sámi society that the traditional Sámi religion does not exist anymore. The symbolism and figures of the drums are therefore obsolete for many. In addition, many Sámi people today are Christianized and reject their ethnic religion.

Other contributing factors regarding the historical background of the Sámi can be understood in a similar fashion. Sámi historian Veli-Pekka Lehtola (2015) outlines the existing problem of denial within the state of Finland concerning the whitewashing of crimes against the Sámi. These include sending Sámi children to residential boarding schools, the theft of land, forced conversion to Christianity and a long history of cultural violence against the Sámi. "And yet, even in the 2000s, there have been strong arguments also among Finnish historians in the 2000s emphasizing that the treatment of the Sámi in Finnish history was equal and fair. It seems to be valid that the historical relations of the Finns with the Sámi have been different from the relations of the Scandinavians to the Sámi, but there has even been argumentation entirely denying the colonialism" (Lehtola, 2015, p. 23).

Johannes Schefferus mentions the drum in his *History of Lapland* (1674). Schefferus describes receiving the drum from "Magnus Gabriel de la Gardie, 'the Chancellor of the Kingdom' of Sweden" (Schefferus, 1674, p. 49). The drum was sent to Sweden and has now been returned to Finland, to the Sámi Siida Museum in Inari, on a permanent loan basis. The drum is one of only two existing drums from the former Kemi Sámi area. The cosmological landscape is divided into three different realms or worlds, which are inhabited by humans, spirits, animals and holy places called *sieidi* and *passe* in Sámi terminology. These symbols and figures are regarded by the Sámi as their common property but are used on packaging and souvenirs in shops throughout Finland (Manker, 1938, p. 686, see Figure 11.1).

As a response to the ongoing exigency regarding the contributing factors, the following subchapters will provide a number of examples and contexts where Sámi cultural heritage has been appropriated within Finnish tourism. In order to

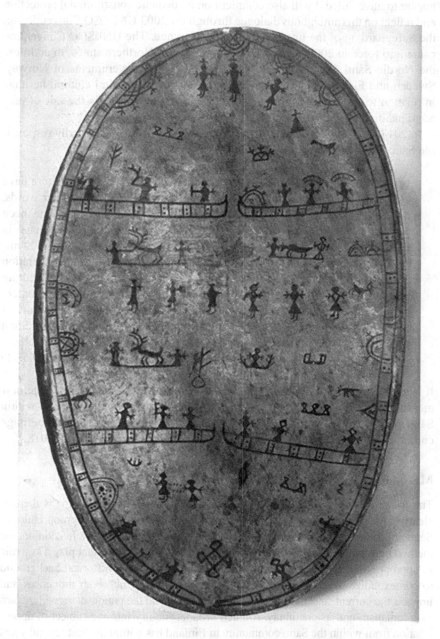

Figure 11.1 A Sámi drum that originated from somewhere within the Kemi Sámi area but has an otherwise uncertain Sámi origin

discuss how subsequent recommendations for the protection of Sámi heritage appear to have failed, I will also comment on inadequate constitutional protection and reflect on this ambiguous dialogue through the 2003 UNESCO Convention on the Safeguarding of the Intangible Cultural Heritage. The UNESCO Convention came into force in 2006 and has been ratified by all Northern states. In addition, the Nordic Sámi Convention of 2017, agreed by the governments of Norway, Sweden and Finland, has not increased the protection of Sámi cultural heritage in relation to appropriation through tourism. This, too, is seen in the sale of fake Sámi handicrafts and representations of culture.

In 2017, the Sámi Parliament in Inari issued guidelines on culturally responsible Sámi tourism. On their website, they addressed the key questions:

> In tourism, utilizing Sáminess in Finland, symbols of the Sámi culture have already been developed into tourism products for decades. In other words, the commodification of the Sámi culture in tourism has, for a long time, been defined as well as realized by outsiders. The commodified Sámi image in tourism utilizing Sáminess rarely has much to do with the authentic Sáminess. The existing primitivised and misleading widely spread representation of the Sámi in tourism utilizing Sáminess in Finland is, at its worst, insulting for and/or commodifying of the Sámi community. This repeatedly presented, public, misleading image negatively affects the vitality of the Sámi culture. It is hard for the Sámi community alone to provide correct and authentic Sámi presentation with limited resources.
>
> (Sámi Parliament, 2017, p. 1)

It is also important to take into consideration recent proposals to introduce new measures to safeguard Sámi culture and traditions that have emerged from within Sámi society in the past four years. These were discussed in a cultural heritage conference I attended at the Sámi Sajos Center in Inari on May 23–24, 2018.

Methodological approach

The inquiry is approached through a descriptive research methodology as there is clearly a need to identify and describe the nature of the problems of appropriation of Sámi collective identity and cultural heritage as a resource in tourism. In addition, the method both describes and brings to light the contributing factors that play a key role in appropriation. For instance, it is critical to understand the historical background to contextualize the study from different time periods, which both influences and impacts the current state of affairs between the Sámi and the nation states with regard to colonialism and its continuity. A much broader inquiry into the subject of appropriation from within the Sámi community in Finland has within the past several years picked up momentum. There is now a much stronger scholarly interest and investigation of the issues of assimilation of Sámi cultural heritage into the tourist industry. This research intends to provide a clear picture of the ways culture is appropriated and to dissect some of the ongoing complexities involved in the discourse as a result.

I have likewise considered it necessary to include photographic data and extracts from interviews with Sámi and Finnish persons, as well as a list of sources (articles) written from within Sámi culture to describe some of the contexts in which appropriation takes place. These help validate the analysis and make the extent of the grievances held by the Sámi better known. The method helps describe the existing conditions between the Finns and the Sámi population and the appropriation of their cultural heritage into tourism and the (ineffective) legislative documents that are supposed to protect the cultural heritage of the Sámi and other indigenous peoples.

Sámi voices speak out against the appropriation of their cultural property

During fieldwork excursions to the municipality of Inari (and Ivalo as its administrative center) in the current Sámi homeland areas in Finland, as well as Muonio in western *Sápmi*, I had become aware of a variety of factory-made drums on sale in shops as a way of making Sámi culture exotic.

The use and reproduction of the worldview of the Sámi concerning their spiritual culture and related symbolism clearly has a negative impact on the traditions. The development of tourism by the Finns reinvents Sámi history and animistic pre-Christian religion in relation to practices that are ethically unsustainable. Sámi cultural expressions are not the same as Finnish cultural expressions, but Sámi cosmology, cultures and traditions are reused in such a way that the economic beneficiaries are the Finns.

Drums and their symbolism are still considered sacred by Sámi people because they are linked to identity, cultural memory and language and are, as a whole, embedded knowledge systems that play a central role and function in the Sámi *Duodji* handicraft tradition.

As a method for opening a dialogue on the Sámi drums and their use and value in contemporary Sámi society, I spoke with Sámi elder and scholar Elina Helander-Renvall on the subject matter (Helander-Renvall, 2018):

> Among the Sami, shaman drums are still treated with respect as they give expression to existence of spirit beings and sacredness of life. In addition, every historical drum and its symbols are still important for the identity, inscape and cohesion of Sami people as they mark out the inner energizing aspects of culture.

To gain further understanding of the reproduction of fake Sámi drums in Finland, I put a second question to Helander-Renvall (2018) concerning Sámi attitudes toward the different contexts drums are used in. I received this response:

> Finns use Sámi shamanism and symbols in their commercial activities as tourist display or selling. Then there are Finns who re-produce and use the "Sámi" drums without commercial purposes: they can be neo-shamans,

Figure 11.2 An artificial representation of a constructed resource for the development of tourism and a marketing tool: two fake Sámi drums exhibited in a souvenir shop in Ivalo

Source: Photograph Francis Joy 2017.

new-agers, healers and people who enjoy using drums and symbols as decoration, gift to friends, etc. We do not know much about this group. Sami follow traditions, as they use the drums. Non-Sami drums and Sami symbols drawn by non-Sami may lack traditional Sami spiritual sensibilities and spirits connected to drums. People usually do not understand what makes a Sami drum spiritually strong.

In relation to what Helander-Renvall states here, important news articles appeared in November 2008, addressing the general misuse and appropriation pertaining to the reuse of Sámi symbolism and cultural heritage within the Finnish tourist industry. The first article by Aslak Paltto (2008) was published by Finland's national public service broadcasting company YLE as "Restrictions on use of Sámi cultural symbols." The second article from K. Mar Hauksson and published in *IceNews – News from the Nordics* under the heading "Finland's Saami protect their cultural symbols" (Hauksson, 2008). These two short articles are important for highlighting the struggle of the Sámi people to save and maintain their cultural heritage and identity; they also help raise awareness of a growing problem in Finland.

Also, voices from outside the Sámi community, such as the Swedish scholar Eva Silvén from the Nordiska Museet in Stockholm, have academically inquired into the use and reproduction of drums from farther afield. Silven (2012) also addresses from within Sámi society some of the conflicts that have arisen due to new kinds of drums made by Sámi persons themselves and their rights to produce such artifacts.

Further, in her master's thesis, titled: the representation of Sámi people on Finnish and Norwegian tourism websites in English (2014), Maiju Lindholm deals with a broad range of issues concerning discrimination and the false representation of Sámi people and their culture within tourism and advertising. More recently, an article appeared in the YLE news (2015) about Sámi people trying to stop the exploitation of indigenous handicrafts.

In an attempt to try and establish the perspectives within the tourism industry in Finland and especially in Lapland with regard to ethical guidelines for the production and use of Sámi costumes, as well as the production and sale of souvenirs, I contacted Rauno Posio from Visit Arctic Europe, which is part of the Finnish Lapland Tourist Board. This is what he had to say (Posio, 2018):

> Ethical guidelines for the use of Sámi heritage were made between the Sámi Parliament and House of Lapland [Lapland's official Marketing and Communications Company], but only concerning the Sámi traditional costume – *Gákti*. There are no guidelines for the production and sale of souvenirs. The guidelines for the costume were recently made in the autumn 2018 because the tourist industry has been making videos and promoting tourism in Lapland through misuse of the Sámi *Gákti* in order to increase business and sales. There is however a growing awareness concerning looking for gifts that are authentic amongst European customers when it comes to buying souvenirs. Yet, the area where most of the profit is made comes from fake *Duodji* or souvenirs that are not authentic. Some businesses in Lapland have stopped selling cheap souvenirs. For example, Arktikum House in Rovaniemi is selling only authentic items but have low profits as a result.

I put a further question to Rauno Posio about whether there were any monitoring systems on what was produced and sold regarding the Sámi costumes and fake souvenirs and whether or not companies who did not follow the ethical guidelines were held accountable. His answer: "No, not at the moment" (Posio, 2018).

What are the mechanisms that are supposed to protect Sámi cultural heritage?

The next subchapter will feature documents that have failed to provide adequate constitutional protection for the cultural heritage of indigenous peoples, in this case the Sámi. Not only has legislation been unable to protect Sámi cultural heritage from unethical tourism practices, but there are also various other complexities.

While proposals have been issued to safeguard the cultural heritage and sustainable tourism practices, policies are actually administered through tourism, which arguably seeks to weaken Sámi culture, traditional religion and traditions through assimilation. This could be seen as violating the rights of the Sámi to control, maintain and govern their traditions and practices, and thus further dispossesses them of protecting their heritage.

For example, the 2003 UNESCO Convention on the Safeguarding of the Intangible Cultural Heritage, which has been ratified by Norway, Sweden and Finland, lays down the following (Article 11):

> Each State Party shall: (a) take the necessary measures to ensure safeguarding of the intangible cultural heritage present in its territory, (b) among the safeguarding measures referred to in Article 2, paragraph 3, identify and define the various elements of the intangible cultural heritage in its territory, with the participation of communities, groups and relevant non-governmental organization.
>
> (UNESCO, 2003, p. 4)

It should likewise be noted that intangible cultural heritage also becomes tangible cultural heritage because, for example, drums and their use are related to oral traditions, *joiking* and performance and represent local knowledge painted on the drumheads.

For the benefit of the reader, Article 2, paragraph 3 defines the following:

> "Safeguarding" means measures aimed at ensuring the viability of the intangible cultural heritage, including the identification, documentation, research, preservation, protection, promotion, enhancement, transmission, particularly through formal and non-formal education, as well as the revitalization of the various aspects of such heritage.
>
> (UNESCO, 2003, p. 2)

The draft for the Nordic Sámi Convention 2017, compiled by the Ministry of Justice, Finland, is a treaty. Yet the document has not been ratified and therefore remains a proposal at this stage. However, in terms of guidelines concerning the protection of Sámi cultural heritage, it is very clear on

> Sámi cultural heritage, expressions of Sámi culture and Sámi traditional knowledge: The States shall take the Sámi cultural heritage, expressions of Sámi culture and Sámi traditional knowledge into account in decision-making concerning Sámi circumstances.
>
> The material cultural heritage of the Sámi shall be protected by law. The States shall promote the protection of the immaterial cultural heritage of the Sámi. The Sámi Parliament or authorities in cooperation with the Sámi Parliament, Sámi museums or other Sámi institutions shall be responsible for issues relating to the Sámi cultural heritage.

The Sámi Parliament, or museums or institutions in cooperation with the Sámi Parliament, shall be responsible for managing Sámi cultural heritage in the custody of the State or another public institution.

The states shall promote the repatriation and management of Sámi cultural property in accordance with the third paragraph.

(Nordic Sámi Convention, 2017, p. 7)

This sounds very good on paper but is as yet ineffective and set to fail in relation to preventing appropriation of Sámi cultural heritage into tourism in Finland because of poor enforcement and a general lack of accountability. The document has not been ratified and is therefore doomed to fail.

Cultural hostage taking and the Sámi struggle for recognition

Assessment of these documents and the developments since the UNESCO 2003 Convention show that the protection of the Sámi cultural heritage has not improved. Appropriation has in fact become more widespread in the wake of developing tourism.

In the case of the Sámi, everything looks good in the moral sense that Finland is moving "towards or considering" the plight of the Sámi in terms of protecting their traditions. Still, the Sámi are and have been for a very long time in what I would refer to as a cultural hostage-taking situation. Quite simply, they are enveloped by the dominant majority group of Finns. Furthermore, the government keeps subordinating the Sámi, appropriating their collective heritage and identity, thus threatening their culture.

These acts of racialized inequality allow the Sámi to fade away. The subtle forms of what could be termed a sustained violation and weakening of the Sámi traditions further assimilate them into Finnish culture. The Sámi are denied legitimate ownership and protection of their culture, or, as Grzanka (2010) puts it:

> Legitimation is similar to Gramsci's (1971) hegemony: it connotes a system of ruling (or even domination) that at least appears to be democratic and inclusive, even if the material reality is quite the opposite.

(Grzanka, 2010, p. 94)

What we have here is a good example of the "confluence of practices, processes, and institutions that reify historical inequities [and serve] to ultimately validate white privilege" (Anderson, 2013, p. 39).

Difficulties of building a framework that protects Sámi cultural heritage

A lack of underlying structures plays a key role in the creation of unsustainable and dysfunctional tourism. In order to gain a broader perspective and better understanding of why there is such a lack – an absence of effective legislation that protects Sámi cultural heritage from assimilation – a conference on Sámi intangible cultural heritage was held at the Sámi *Sajos* Center, Inari, on May 23–24, 2018. The center, where the Sámi Parliament also convenes, hosted representatives of the UNESCO; the Sámi Parliaments of Norway, Sweden and Finland; the Sámi Council from Norway and members of the Russian Sámi community from the Murmansk area on the Kola Peninsula in Northwest Russia.

The lectures were mostly presented in Sámi, Norwegian, Finnish and Russian languages and were subsequently translated by translators working at the Parliament. I compiled a summary of all the translated presentations to further elaborate on why there is an interactive web of exploitation regarding the appropriation of Sámi cultural heritage into tourism in Finland. This is in order to show that there are also issues within the Sámi community itself which contribute to the lack of protection and subsequent assimilation of Sámi heritage.

In her presentation, Finnish researcher Kirsi Suomi tackled the misuse of Sámi culture in Finnish tourism, arguing that, in order for legislation to be effective in protecting cultural heritage, "collective decisions between Sámi Parliaments are needed, in relation to cultural identity theft, cultural appropriation, invented traditions – traditions exploited; borrowed traditions, stereotypes, exoticization, primitivisation and zooification" (Suomi, 2018). Her argument reflects what was highlighted by the earlier photographic examples regarding fake drums that are invented and modified to make a profit while exploiting Sámi traditions and religion.

Ellen Berit Dahlbakk from the Sámi Council discussed the Sámi *Duodji* trademark and how it had been difficult to "implement ethical rules inside the Sámi community because Sámi artists themselves are using designs from other parts of Sápmi, for their own innovations, which is problematic" (Dahlbakk, 2018). Her main points resonate with what was said earlier and foreground the complexity involved in addressing these issues within the Sámi community. They also echo the notions by Gunvor Guttorm regarding fake *Duodji*, where Sámi artists ignore the guidelines for reproducing traditional knowledge. However, the fact that Sámi artists reuse drum symbolism and cultural heritage freely also shows growing concerns about reductionism and the weakening of Sámi identity.

The representative of the Sámi Parliament in Finland, Anni-Helena Ruotsala, examined the ethical regulations and the understanding of the intangible cultural heritage, stressing that "a greater concerted effort is needed to define what effective measures are needed to protect cultural heritage" (Ruotsala, 2018). Moreover, she showed what has evolved from within the Sámi Parliament regarding academic study: researchers now need a permit to undertake research into Sámi culture. In terms of scientific study, this is problematic because it cannot be guaranteed that everyone will follow these requirements as they will mean a radical change in planning, supervision and heritage management. The situation will be equally complex within the tourism industry, which will need to resolve how culture is approached and reproduced by both Sámi and non-Sámi organizations and actors. It will take much effort to reinforce the shared values and beliefs within Sámi culture and to ensure their protection and preservation.

Pia Nuorgam, a Sámi researcher from the University of Lapland, gave her presentation on Sámi culture and evolving copyrights. According to her, there is "lots of talk about the importance of cultural heritage but there have been no concrete decisions" (Nuorgam, 2018). This also served as a timely reminder in terms of what was discussed earlier in relation to examples of failed legislation and weak laws that do not protect cultural heritage. More importantly, it also brings into question whether or not the Sámi Parliament in Finland, for example, has the resources to effectively manage cultural heritage and decision-making when it comes to protecting and maintaining the fabric of Sámi society, whose identity is deeply rooted in both the past and the present. Moreover, what kind of plans or schemes does Parliament have for stopping the reproduction of fake *Duodji* within the tourism industry with regard to protecting and retaining copyright?

What are the consequences of using Sámi cultural heritage as a resource in tourism?

The impacts of tourism add to impoverished Sámi traditions and culture by non-Sámi actors because the concept, origin, usage and symbolism of sacred drums are downgraded in cheap replicas, which further reduces Sámi traditions and values. Moreover, the reinvention and recontextualization of drums as artistic objects that have undergone extensive modification in order to reproduce a false representation of Sámi cultural heritage for consumerist purposes exemplify the harm caused by a lack of sustainable and responsible laws and legislation. I use the term "false" because original drums were made to be spiritually strong and were consecrated for the purposes for which they were made (healing, divination, trance). This is not the case with drums that are factory made. Therefore, one could argue that harm is caused because the information misleads tourists who buy them, and the Sámi Parliament has not authorized the creation and sale of fake drums.

From within academic tourism studies in Norway, Trude Fonneland (2017) discusses indigenous spirituality and the processes of cultural branding in the context of tourism. She refers to the works of Petterson and Viken (2007, p. 185), who note that "developing commercial products out of important culturally based spiritual icons was (in Sámi society) seen to be morally questionable, and it certainly impacted authenticity" (Fonneland, 2017, p. 256).

As regards Finnish assimilation policies in relation to the Finns representing Sámi heritage through dysfunctional tourism practices, it can be argued that tourism creates a division of both wealth and power in economic development and cultural sustainability. Moreover, sectors of the industry create a false and diluted portrait of the Sámi religion, making it understandable why the Sámi have little faith in negotiations with the Finnish government concerning the protection of cultural heritage. Tourism studies also highlight one of the main problems around

Figure 11.3 A plastic tray containing a combination of drum landscapes that have been copied and modified from the sacred *noaidi* drum from Finland. The tray was on display in a souvenir shop in Ivalo in Finnish Sápmi. These illustrations are sources of traditional knowledge for the Sámi and are therefore considered their common property. These designs also appear on wallpaper and curtains

Source: Photograph: Francis Joy 2017.

assimilation of cultural heritage into tourism in Finland. As Seija Tuulentie contends in her analysis (2017) on whose voice is heard and how, visitors often fail to understand that, when it comes to business development in Lapland, "community has been understood as one entity" (Tuulentie, 2017, p. 122). The Sámi are typically seen as conquered and absorbed into mainstream Finnish culture – and the reproduction of Sámi heritage by the Finns is a symbolic reference to this very understanding. Addressing these issues would therefore present a major challenge to the tourist industry in relation to the Sámi recovering their heritage, identity and self-determination.

The repatriation and revitalization of the Sámi was also explored by Eeva-Kristiina Harlin at the 2018 conference in Inari. She discussed the ways in which "the Sámi cosmology changed when the Sámi children were sent to boarding schools" (Harlin, 2018), which relates to what I have noted about the outlawing of the Sámi religion. There is thus a belief or comprehension that the Sámi religion no longer exists, as it has been assimilated into mainstream culture. This diminishes the value of the symbolism on the drums, which is related to the language, belief and pre-Christian religion that are all tied with cosmology. Sámi children were not allowed to practice their ethnic religion because it was considered to be associated with evil and the devil. Many facets of the ancient religion were replaced by or combined with Christianity, often with aspects of the Laestadian branch of the Evangelical Lutheran Church.

For the most part, it seems clear that the tourism industry in Finland does not understand that the drums and their symbolism have and always will be a source of livelihood for the Sámi people. The drums matter, because they contain the Sámis' language, memory, cultural landscapes, hunting rituals, waterscapes and perceptions of the world and its inhabitants.

Concluding remarks

A romantic portrait of the history, culture and religion of the Sámi people is used as a retailing tool that designates Lapland as a fabulous tourist destination. It presents a highly ambiguous understanding of culture. Appropriation of Sámi cultural heritage into tourism does not promote the sustainability of Sámi traditions or the vitality of Sámi culture. Instead, it adds to the demise and decline of both culture and traditions because they are often represented by others than Sámi people themselves and because they are poorly defined and reduced for marketing purposes. While tourists and, indeed, many Sámi and Finns see the drums and their symbolism as beautiful, mystical and unique – and as promoting business development and the tourism enterprise in the Arctic North – these practices in fact corrode, eat away at and reduce culture.

It is difficult to create cultural sustainability through tourism development in Lapland with regard to the souvenir industry, as the production and marketing practices are in a state of disarray. There are guidelines for sustainability, but they are purely voluntary by nature and do not amount to legislation. What are the benefits, then, of having them? If there is no law, the guidelines cannot be enforced.

Numerous tourism networks operate throughout Lapland, some of which follow the guidelines while others do not, thus highlighting different layers of complexity. Also, contrary to what has been mentioned in Finland, it is inappropriate to suggest that the excessive marketing of Sámi traditions through tourism is a way of aiding heritage conservation.

"Cultural sensitivity" is a new term introduced into tourism and marketing recently in Lapland. It brings together the concepts of blurred shared values and adherence to ethical guidelines as a symbol and representation of responsible tourism practices. Cultural sensitivity might be seen as contributing to a more sustainable and fairer way of business development, but there is an element of illusion involved in this. As the Sámi handicrafter Neeta Jääskö argues, cultural sustainability cannot be achieved as long as "[t]he exploitation of religious freedom takes place through the appropriation of religious symbolism. Sámi religion is a living tradition that has value, but cheap products have no real meaning" (Jääskö, 2018). There is a clear distinction between the authentic and the counterfeit – but it is a distinction that is typically not clear to the consumer.

References

Anderson, A. (2013) "Teacher for America and the dangers of deficit thinking," *Critical Education*, 4(11) [Online]. Available at: www.academia.edu/31760571/Teach_For_America_and_the_Dangers_of_Deficit_Thinking (Accessed: February 8, 2019).

Bydler, C. (2017) "Decolonial or creolized commons? Sámi duodji in the expanded field" in Aamold, S., Haugdal, E. and Jørgensen, U.A. (eds.) *Sámi art and aesthetics: contemporary perspectives*. Aarhus: Aarhus University Press, pp. 141–158.

De Jong, C. (2017) *Sámi identity, representation and revitalization in northern Norway*. MA thesis, Leiden University [Online]. Available at: https://openaccess.leidenuniv.nl/bitstream/handle/1887/56824/Jong%2C%20Charlotte%20de-s1134396-MA%20Thesis%20CAOS-2017.pdf?sequence=1 (Accessed: February 8, 2019).

Fonneland, T. (2017) "Sami tourism in northern Norway: indigenous spirituality and processes of cultural branding" in Viken, A. and Müller, D.K. (eds.) *Tourism and indignity in the Arctic*. Bristol: Channel View Publications, pp. 246–260.

Gramsci, A. (1971) *Selections from the Prison Notebooks (SPN)*, edited and translated by Hoare, Q. and Nowell-Smith, G. New York: International Publishers.

Grzanka, P.R. (2010) *White guilt: race, gender, sexuality and emergent racisms in the contemporary United States*. PhD dissertation, University of Maryland [Online]. Available at: www.researchgate.net/publication/277228626_White_Guilt_Race_Gender_Sexuality_and_Emergent_Racisms_in_the_Contemporary_United_States (Accessed: February 20, 2019).

Guttorm, G. (2007) "Duodji – Sami handicrafts – who owns the knowledge and the works?" in Solbakk, J.T. (ed.) *Traditional knowledge and copyright*. Karasjok: Sámikopiija, pp. 61–94.

Hauksson, K.M. (2008) "Finland's Saami protect their cultural symbols," *IceNews – News from the Nordics* [Online]. Available at: www.icenews.is/2008/11/18/finland%E2%80%99s-saami-protect-their-cultural-symbols/#axzz4t6wO4Pb4 (Accessed: November 1, 2017).

Helander-Renvall, E. (2018) "Sámi drums, spirits and their roles," in Joy, F. (ed.) *Sámi shamanism, cosmology and art as systems of embedded knowledge*. Acta Universitatis Lapponiensis 367. Turenki: Hansaprint Oy, p. 193.

Hyltén-Cavallius, C. (2014). "Folkdräkt, kulturarv, svenskhet. Omstridda och särskiljande begrepp" *i: Dressing Swedish. From Hazelius to Salander*. Botkyrka: Mångkulturellt centrum, pp. 23–31. Available at: http://www.konsthantverkisverige.se/1/pdf/konsthantverk-i-sverige-del-1_02.pdf

Lehtola, V.P. (2015) "Sámi histories, colonialism, and Finland," *Arctic Anthropology*, 52(2), pp. 22–36.

Lindholm, M. (2014) *The representation of Sámi people on Finnish and Norwegian tourism websites in English*. Master's thesis, University of Jyväskylä [Online]. Available at: https://jyx.jyu.fi/dspace/bitstream/handle/123456789/44816/URN:NBN:fi:jyu201412053434.pdf?sequence=1> (Accessed: September 19, 2017).

Manker, E.M. (1938) *Die lappische Zaubertrommel. Eine ethnologische Monographie. 1, Die Trommel als denkmal materieller Kultur*. Stockholm: Thule.

Mathisen, S.R. (2010) "Indigenous spirituality in the touristic borderzone: virtual performances of Sámi shamanism in Sápmi park," *Temenos: The Finnish Society for the Study of Religion*, 46(1) [Online]. Available at: www.researchgate.net/publication/297561688_Indigenous_Spirituality_in_the_Touristic_Borderzone_Virtual_Performances_of_Sami_Shamanism_in_Sapmi_Park (Accessed: February 10, 2019).

Ministry of Justice, Finland (2017) *The draft for the Nordic Sámi convention*, English translation. Helsinki: The Governments of Finland, Norway and Sweden.

Paltto, A. (2008) *Restrictions on use of Sami cultural symbols* [Online]. Available at: https://yle.fi/uutiset/osasto/news/restrictions_on_use_of_sami_cultural_symbols/6117478 (Accessed: September 1, 2017).

Petterson, R. and Viken, A. (2007) "Sami perspectives on indigenous tourism in northern Europe: commerce or cultural development" in Butler, R. and Hinch, T. (eds.) *Tourism and indigenous people: issues and implications*. Oxford: Butterworth-Heinemann, pp. 176–187.

Sámi Parliament of Finland (2017) *Culturally responsible Sámi tourism* [Online]. Available at: www.samediggi.fi/ongoing-projects/culturally-responsible-sami-tourism/?lang=en (Accessed: November 29, 2018).

Schefferus, J. (1971, 1674) *The history of Lapland, wherein are shewed the original, manners, habits & customs of that people*, Facsimile edn. Stockholm: Bokforlaget Rediviva.

Silven, E. (2012) "Contested heritage: drums and *sieidis* on the move" in Poulot, D., Guiral, J.M.L. and Bondenstein, F. (eds.) *National museums and the negotiation of difficult pasts*. Conference proceedings from EuNaMuS, Identity Politics, the Uses of the Past and the European Citizen, Brussels, January 26–27. Linköping: Linköping University Electronic Press [Online]. Available at: www.ep.liu.se/ecp/082/ecp12082.pdf (Accessed: March 20, 2017).

Sunna, H. (2006) *Duodji – Árbi Arvet: handicraft in the Sámi culture*. Luleå: Luleå Grafiska.

Tuulentie, S. (2017) "Destination development in the middle of the Sápmi: whose voice is heard and how?" in Viken, A. and Müller, D.K. (eds.) *Tourism and indignity in the Arctic*. Bristol: Channel View Publications, pp. 122–136.

UNESCO (2003) *Text of the convention for the safeguarding of the intangible cultural heritage* [Online]. Available at: https://ich.unesco.org/en/convention#art11 (Accessed: January 24, 2018).

Zickerman, L. (1918). Äktenskapets ekonomi. *Hemslöjdkommitténs betänkande*, p. 230.

Lecture notes

Dahlbakk, E.B. (2018) *Work in progress: the Sámi duodji trademark*. Inari: Sajos Cultural Centre.

Harlin, E.K. (2018) *Repatriation and revitalization*. Inari: Sajos Cultural Centre.

Jääskö, N. (2018) *Sámi duodji. Women's entrepreneurship and cultural sensitivity in tourism*. Seminar, University of Lapland, November 29.

Nuorgam, P. (2018) *Sámi culture and evolving copyrights*. Inari: Sajos Cultural Centre.

Ruotsala, A.H. (2018) *Ethical regulations and the understanding of the intangible cultural heritage brought for the outside professionals*. Inari: Sajos Cultural Centre.

Suomi, K. (2018) *Misuse of the Sámi culture in Finnish tourism*. Inari: Sajos Cultural Centre.

Telephone interviews

Posio, R. (2018) *Ethical guidelines for sustainable tourism in Lapland, personal communication*.

12 History as a resource in Russian Arctic politics

Susanna Pirnes

Introduction

Russia is the largest Arctic state in terms of its size, sea boundaries and interests in the region. The collapse of the Soviet Union left the Russian Federation in a limbo of a state lacking national identity. The Arctic is a good example of region building, paralleled in the Soviet project of "conquering" the North in the 1930s. The early years of the Soviet Union defined Soviet identity on the basis of boundless space, and unpopulated areas in the North were conquered to modernize the whole of the Soviet Union, including the backward countryside and the white spots on the map. These modernization processes were to prove that nothing was impossible – that even nature was controlled by the new Soviet man.

Mastery of space has been an important aspect in the formation of Russia and in the development of its territories far to the East, South and North and the cosmos. The memory of its colonial history still defines the self-understanding of Russia as an imperial state, so much so that the Arctic region as a space and its historicity are an inherent part of the discussion in Russian philosophy about infinite space and its relation to the Russian mentality. Philosopher Nikolai Berdâev (Berdyaev) (1874–1948) developed the idea of Russia's longing for an unbounded space and how this influenced both the mind-set and the governance of the vast country.

The Arctic has become a hotly debated subject, especially since the breaking of the news that it is warming at twice the rate of the rest of the globe. A new kind of Arctic has been in creation from the 1990s in the utopian narratives about the (non)place (Dahl, 2012). Those narratives often refer to the heroic past of conquering the North, and Arctic states are taking part in this process by commemorating their Arctic histories (Eng and Lindaräng, 2013). This is seen in their national Arctic strategies, the launch of different programs to enhance the development in the North and new claims about the Northern territories.

Anssi Paasi has defined the process of the institutionalization of a region as having four consecutive stages: 1) geographical region, 2) symbolic appearance, 3) regional organizations and 4) identity and regional consciousness (Paasi, 2002, p. 140, referred to in Riukulehto, 2015, p. 42). In the Russian context, the Russian Arctic zone was defined by a presidential decree in May 2014, so the geographical region and its symbolic appearance exist. Regional organizations have been established, such as the State Commission on the Arctic (2015), the Arctic Commission of St. Petersburg (2018), the Northern Sea Route Administration (2013) and various universities and research institutions.

Russia is now strengthening its Northern identity on many political and regional levels. I will analyze this process of (Arctic) identity building more closely within the framework of historical culture. By historical culture, I refer to those encounters where people meet their past (Ahonen, 1998), be they monuments, buildings, architecture, city planning, landscape, museums, popular culture or commemorative practices. Through historical culture, the past is present in a concrete manner in individuals' and communities' lives, so it can be understood as a space where they construct the relationship to the past and the future (Tilli, 2009, p. 280). Historical culture is an integral part of a living environment; it both reflects ongoing changes in the North and can be interpreted as a symbol of that which is constant. Questions of social and cultural sustainability have become more important in the Arctic when the "constantly changing" Arctic and its people need to be more resilient to overcome the changes. At the same time, social and cultural change can be seen as the normal state culturally, as there have been communities and cultures in the Arctic for thousands of years (Lempinen, 2018).

Memory and history are often understood as synonymous but must be differentiated at least for methodological purposes. By adding a third dimension of time, we get a triangle in which the edges are in constant interaction (Nyyssönen and Halla, 2013, p. 24). It is obvious that the past and the present interact in the memory processes, but we should not forget the future either. Our memories help us build scenarios with good or bad outcomes, which makes these future memories just as important as those of the past (Ned Lebow, 2008, p. 39).

I will examine some observations from the last ten years that represent the presence of the great past in the new narrative about the Great Russian Arctic. I refer here to the official speeches and statements of Russian leaders, who tend to present the Arctic as a "historically important" part of Russia. I will also analyze a photographic exhibition representing the "official" Arctic at the time of the opening of the International Polar Year in 2007. My main question is how the Russian Arctic is portrayed in historical culture and how these representations reflect Russian Arctic politics in the light of identity formation.

Marlene Laruelle (2014, pp. 9–10) points out that in the Kremlin's search for political legitimacy through strong patriotism, the Arctic has been assigned a flagship role in nationhood building. I take her claim as a starting point to show how the remembering, presenting and commemorating of Arctic history has helped bring the Arctic back into the Russian national narrative during the last decade.

Russian Arctic politics

The planting of the Russian flag on the seabed at the North Pole in 2007 is commonly said to have marked the re-energizing of Russian Arctic policy, but policy operations had started well before this date. In 2001, Russia had already become the first nation to deliver its claim to the extended continental shelf under the United Nations Convention on the Law of the Sea (UNCLOS). But the 2007 events represented a symbolic turn and more: in the words of Artur Čilingarov, polar explorer and president of the State Polar Academy, "the North is ours" (MK.ru, 2007). Russian Arctic policies are typically interpreted as expansive and aggressive in the West, and statements such as Čilingarov's do not conflict with these interpretations (see e.g. Sergunin and Konyshev, 2014).

The process in Russia of developing the Arctic region and strengthening national security started at the beginning of Vladimir Putin's first presidential term. As mentioned, Russia was the first nation to deliver its claim to the region, including a claim to the continental shelf, to the UN's relevant tribunal as early as 2001. In 2004, a memorandum on the development of northern regions was published by Russia's state council (*Gosudarstvennyj sovet*), an advisory body to the Russian head of state, consisting of the leaders (governors and presidents) of Russia's federal subjects. This memo already listed themes prominent in the actual Arctic strategy published later (2008 and 2013), and in 2004, Russia's role in the global economy was already said to be largely defined by the Northern region. Russia's Arctic region has certainly been a part of President Putin's identity politics, but the roots go further back in time, to the propaganda value of the heroic Arctic stories of the 1930s (Pirnes, 2017, p. 68).

Russian military activity in the Arctic has been growing for years. While new military bases have also been in the making, it has mostly been about reopening old Soviet bases. A new one, Russia's northernmost military base, in Franz Josef Land, is said to be the largest building in the circumpolar Arctic (Barents Observer, 2017).

The Northern Sea Route and the significant potential attached to this Russian segment of the Northeast Passage have been debated long and hard. Statistics suggest that the potential of the sea route has been exaggerated, also in light of the underdeveloped infrastructure and inadequate search and rescue preparedness. President Putin has set a goal of 80 million tons of cargo annually to be transported along this route by 2024. The Arctic Commission was reorganized at the end of 2018, with a new chair and fewer members in order to speed up the decision-making and to make it more flexible. The new commission is tasked with ensuring that the president's order on the growth of Northern Sea Route traffic is met (Barents Observer, 2018).

Russia's claims to the continental shelf can also be seen as a natural continuation of the expansive Soviet policies in the Arctic, especially in relation to the sectoral policy, which maintained that the sectors drawn from the most Western and Eastern tips of the Soviet Union to the North Pole clearly defined the sea area under Soviet control. The sectoral policy draws on Canadian senator P. Poirier's idea to divide Arctic sea areas to sectors between nations holding Arctic coastlines. Poirier presented the concept to the Canadian Senate as early as 1907. Canada eventually accepted the proposition in 1925 in an amendment that declared Canada's sovereignty all the way to the North Pole. The following year (1926), the Soviet executory central committee published a decree referring to the sectoral model and claiming all land masses and islands within their sector as Soviet property (Timtchenko, 1997). In this perspective, the claims that Russia has brought to the UN uphold the sectoral principle and are based on the historical experience of changing attitudes toward the sector theory during the 1930s and 1940s (Horensma, 1991, pp. 72–80).

History as a resource

The collapse of the Soviet Union left the Russian Federation in a state of uneasy national identity. Karaganov et al. (2014) studied Russian national identity twenty years later. They argue that the time is now ripe for a rebirth of Russian national identity, as the remnants of Soviet identity are dying, and a pre-revolutionary Russian identity has not yet been reestablished and cannot be reestablished in its full form.

The Russian Federation is a successor state of the Soviet Union, so the territorial and identity losses were enormous. History has been used to mend the deficits in various ways in Russia during the last two decades. Especially striking has been the use of the memory of the Great Patriotic War. The "right" interpretation of history has been regulated even with the help of a Presidential Commission of the Russian Federation to Counter Attempts to Falsify History to the Detriment of Russia's Interests (in place in 2009–2012).

Given this background, the history of the Russian Arctic is a potential resource for identity building. Historical events are being remembered in many ways, depending on personal histories and knowledge about the past. The most visible and obvious objects are monuments and statues, which present a certain interpretation of the past. The relationship of the people and their (nation's) past is negotiated on the level of historical culture. The determination to strengthen Russia's Arctic identity is related to the economic and security interests of the state but is also a continuation of the complex relationship of Russian identity to space (cf. Mjør, 2017). For the authorities, identity building legitimizes Russia's huge and expensive construction and development projects, and for the Russians, it can be interpreted as guaranteeing Russia's greatness and boundlessness – the fact that Russia remains a superpower.

While the historical culture of the Arctic has not been the focus of interest in Arctic research, Arctic (natural) resources and related aspects have been very popular topics in academia, especially as regards sustainable development and social sustainabilities. History is still often perceived as a chronicle of events that have led to the present situation. By challenging the Rankean perception of "objective history" and by foregrounding history, it is possible to link history to the discussions of social and cultural sustainabilities in the Arctic. By cultural sustainability, I understand a set of practices and perceptions that are aimed at sustaining culture. Interestingly, the official strategies of Arctic states address culture only in the context of indigenous cultures (Lempinen and Heininen, 2017, p. 5). The historical culture of the Russian Arctic most often represents the Russian view on history from the vantage point of the (ethnic Russian) political elite.

Arctic as space

The Russian Arctic is a hybrid term referring to areas which are discussed as having no clear borders or definitions. The Arctic zone of the Russian Federation was defined in 2014 so the state could take a tighter grip on these often very remote areas. The Russian North is a far more open-ended term than the Russian Arctic in that the Russian Arctic is seen as part of the Russian North (Fedorov, 2018). It is not meaningless to ponder these concepts as they have so much weight in Russian culture.

The North as a concept (not as a compass point) refers to, among other things, to the "Russian North," which is usually understood as the northern parts of European Russia settled by ethnic Russians. Russian North has a strong symbolic meaning: Russian Northern culture is considered to have preserved the purest form of traditional Russian folklore. There are voices that demand that more attention be paid to the Russian North, instead of concentrating all the energies on the geopolitical construct of the Arctic zone (Šabaev et al., 2016, pp. 129, 133).

Emma Widdis (2003, pp. 3–5) has studied the meaning of space and territory in Russian history and in relation to the construction of Sovietness. She refers to the huge mapmaking processes in the early Soviet period, which aimed to create a coherent "imaginary cartography" for the sake of a new Soviet identity. She claims that it still has a profound significance for the post-Soviet identities. Space is hardly ever a mere space (Rogachevski and Steinholt, 2017, p. 1). Mapmakers' intentions, worldviews and cultural values and the semiotic subsystem affect the mapmaking, but they might not have the intended impact. As a result, Soviet people constructed mental maps that were more accurate than the printed maps (Borén, 2009, pp. 175, 197, referred to in Rogachevski and Steinholt, 2017, p. 1).

How has the idea of space been implemented in politics or used to legitimize Russian Arctic politics? Sergei Medvedev (2018) writes that "the symbolic significance of the Arctic has far prevailed over its practical nature" and claims that "in fact, Russia took pride in possessing one immense emptiness." As the Russian North could not provide any typical colonial wealth to Russia, the Arctic was more a symbolic asset to Russian sovereignty and security, Medvedev argues. His point of view is certainly that of a person who does not live in the North himself. His is a "typical colonial" attitude toward the North; rather than taking into account the indigenous cultures, Medvedev claims the North is an empty space. (Stammler, 2019.)

The centralized development in Arctic region building is also seen in the process of new comprehensive legislation, which will cover the entire Russian Arctic zone. This legislation is to be accepted in 2019 – despite mounting criticism. The process of tightening up the administration of vast areas, which belong to different regions and oblasts, exemplifies Russia's historical center-periphery (Moscow vs. other areas) relations. There was a period in the history of the Russian North when a huge state organ, Glavsevmorput (*Glavnoe upravlenie Severnogo morskogo puti*), called a state within a state, was responsible for Arctic research, shipping, mineral production, shipyards and aviation, with 100,000 employees

(Laruelle, 2014, pp. 26–27). Soon after the establishment of the Glavsevmorput office in 1932, its powers were centralized in Moscow. There is a current parallel in that Vladimir Putin has little by little taken power away from the regions after the liberal 1990s, when President Yeltsin granted them a lot of autonomy.

Institutional use of history

Victorious history is commonly harnessed to amplify the chosen political path and to fortify the preferred historical narrative. Historical culture often has a major role in the process. Not only are victories and historical success stories rehearsed, but they are also intertwined with traumas and tragedies. It is by no means rare to use traumatic events from the past to stir up popular campaigns that, in one way or another, aim to cope with the difficult past. Usually the actors in these processes are political elites who seek to institutionalize the desired interpretation of history.

While history is full of tragic events, the Russian historical culture of the Arctic focuses on the positive or victorious outcomes. For example, the recommendations for a nationwide school lesson on the Arctic as a facade of Russia, issued in 2016, stress the longstanding history of exploration, which started in the 11th and 12th centuries by the Pomors and Cossacks (Metodičeskie, 2016). According to the recommendations:

> The history of the openings, research and conquest of the Arctic is full of bright, meaningful, at times dramatic events. Knowledge about and awareness of these events can further the education of pupils by forming in them a feeling of pride about the history of research and territorial conquests of their country.
>
> (Metodičeskie, 2016, p. 2 transl. SP)

When talking about the Russian North, history politics particularly translates into paying homage to well-known Russian polar explorers, such as Georgij Sedov and Ivan Papanin; to aviator Valery Čkalov, who pioneered the polar air route from Europe to the American Pacific Coast; to Operation Derviš as the first of the Allied convoys in aid of the Soviet Union in WW II (Dervish); to great technological triumphs such as icebreakers and to monuments. History politics also means remembering the Gulag forced-labor camps. All these examples represent an institutional use of history. Certain aspects of an event or a site are commemorated, such as the Gulag, but only partly. In popular culture, the history of "fulfilling the plan" is being retold and reinterpreted, especially the camp system in the Northern areas. Young authors, such as Guzel Yakhina (2019) in her novel *Zuleikha Opens Her Eyes*, tell the other side of the great, victorious "conquest" of the Arctic by focusing on the inhabiting of uninhabited Siberian lands by unending flows of kulaks and their families. Or like best-selling author Eugene Vodolazkin (2018) in *The Aviator* unravels the story of the Solovki prison camp from the perspective of the modern world, depicting the harsh conditions in a matter-of-fact language.

In Russian Arctic politics, there are several institutions that date back to the Soviet era or even imperial Russia. One of the more influential players in Putin's inner circle is the Russian Geographical Society (*Russkoe geografičeskoe obšestvo*). Founded in 1845 by a decree from Emperor Nicholas I, the society has seen its second coming during the Putin era; Putin also chairs the board of the society. Some of the wealthiest Russian "oligarchs" act as supporting members of the society and actively promote politically correct "patriotic" Arctic research (Baev, 2013). The society is a strong advocate of Arctic issues such as the importance of Arctic nature and a proponent of environmental protection zones in the North.

Science policy is another aspect of Russia's active means to use narratives based on the historical experience of "conquering the North" by the Russian Empire and the Soviet Union. Natural sciences have dominated the agenda since the beginning of the 20th century, when, for practical reasons, hydrometeorological research started to gain momentum with a strong emphasis on ice predictions. Many related institutions are located in St. Petersburg, such as the noted Arctic and Antarctic Research Institute, which announces on its website that it represents and carries on the 185 years of history that the Russian hydrometeorological service has accumulated. The achievements of Arctic science are praised on many occasions, and the past successes serve to stress the importance and legitimacy of institutions.

Old new historical culture of the Arctic?

Traces of Soviet-style memory culture, which included the element of hailing certain professions and their heroic achievements, can still be observed in Arctic historical culture. For example, as a result of coining new traditions, the Russian Federation has celebrated Polar Explorer Day (May 21) since 2013. The chosen day marks the start of the first Russian Arctic expedition on the drifting ice station North Pole in 1937. This day was originally promoted by Artur Čilingarov, special representative of the Russian president in the Arctic. A well-known scientist and polar explorer, Čilingarov took part in the historical cruise to the North Pole in 1977 and was there in 2007 when the Russian flag was planted on the seabed under the North Pole. Čilingarov represents the Association of Polar Explorers, which took the initiative to make official a specific day for polar explorers. The association also proposed including an award dedicated to polar explorers in the Russian system of national awards. They justified the proposition as follows:

People who have dedicated their lives to work in the critically difficult conditions of the polar regions, regardless of their specialization, should be noted at the state level for their services to the people and the state. Therefore, the introduction of the title "Honored Polar Explorer" in the system of state awards of the Russian Federation is considered necessary. In addition, the implementation of this initiative *will contribute to the dissemination of patriotic culture, the promotion of historical values among young people, and to the promotion of activity and enthusiasm in the development of the Arctic zone of the Russian Federation.*

(Association of Polar Explorers, transl. *SP*)

The Russian Arctic shows how the public history culture is negotiated between the elites and the public. Many monuments are initiated by the public, but the need to commemorate heroic and tragic events locally is obviously a consequence of the macro-level politics of the Arctic. A case in point is the St. Petersburg monument dedicated to all polar explorers: it was initiated by the chair of the Association of Polar Explorers in St. Petersburg and was funded by private money (Komsomolskaâ Pravda, 2017).

Official memorials, monuments and museums play a unique role in the creation of national identity because they reflect how the political elites choose to represent the nation publicly (Forest and Johnson, 2002). During the Soviet era, public historical culture was permeated by the dominant ideological current of socialist realism, which also marks the monuments and memorials of that era.

The city of Murmansk in the northwest of Russia hosts plenty of monuments dedicated to a "Northern" or "Arctic" topic. The Monument to the Frontier Guards of Arctic was erected along a small alley in the city center in May 2013. The monument depicts three soldiers, the border guards who protected the lives of Murmansk inhabitants against the enemy in the years of the Great Patriotic War. The statue does not deviate from the Soviet-style war monuments, but the moment of its erection is notable, as it coincides with increasing Russian activity in the Arctic.

Visitors to the monument have left comments on a travel website (TripAdvisor), many of them thankful for the border guards' service in the name of the fatherland and for guarding their home city. The monument is covered with flowers on national Frontier Guards Day, testifying to the strong local dimension of this memorial dedicated to the war history of the Arctic city, but it can also be seen as part of the official narrative about the glorious war achievements of Russia (the Soviet Union). These different layers of historical culture – local and national – do not exclude each other, but exist all in one, which confirms the existence of old and new historical cultures.

St. Petersburg is considered a "Northern capital." This usually refers to its history as the capital city of Russia starting from the early 18th century, when Peter the Great moved the capital there from Moscow. The historical legacy probably inspired the idea to nominate St. Petersburg to be "Russia's Arctic capital." It is also a part of the Arctic game; a commission on Arctic issues was established to work under city administration in 2018 (Komitet, 2018).

Aforementioned monument to polar explorer was erected in 2017 in front of the Arctic and Antarctic Research Institute in St. Petersburg. A great towering figure represents a (male) polar explorer with his dogs, a strong man with his best friends. The anonymous, somewhat mysterious figure adds to (re)constructing the exoticism and masculinity of the profession.

Homogenous, masculine and heroic Arctic or something else?

The International Polar Year (IPY) is a tradition that dates back to the 19th century. The first IPY was held in 1882–1883, the second in 1932–1933, the third in 1957–1958 and the latest in 2007–2008.

The imagery of Arctic science typically portrays the Arctic in a stereotypical manner, as a distant, dangerous place where brave men do great things. A thought-provoking exhibition known as the "Arctic!!!" was organized by the Moscow House of Photography to celebrate or commemorate Russian Arctic science, polar explorers and the exoticism of the Arctic. The exhibition was also a co-creation with the Russian Polar Association and private collections, and the opening was timed to coincide with the beginning of the International Polar Year in 2007 (Catalogue, 2007). The aim of this exhibition was to show

> the history of exploring and studying the Arctic and the Russian North. Works by Russian and Foreign photographers depict the images of courageous explorers, seamen, scientists, merchants and politicians, discovering and settling the austere polar region.
>
> (Catalogue, 2007, p. 5)

In addition, the exhibition showcased exoticized photos of indigenous peoples and cultures. Both the exhibition and the motifs of monuments in Arctic cities are imbued with heroic and masculine imagery, creating certain kinds of mental images of the Arctic and the great deeds which man is able to do in this relentless environment. While masculinity prevails, some "softer" topics refer to the roles of women or animals in the Arctic. Women in general are present in a minority of the photographs. Indigenous women are captured at everyday practices, whereas Russian (non-indigenous) women appear as something of a curiosity, distanced from the everyday Arctic. Such photographs portray a beautiful woman in a sunny, snowy (Arctic) landscape or women waking up in a warmed-up tent, one of the women gazing at herself in the mirror. In a third picture, we do not even get to see a real woman: a man working on the drifting ice station North Pole 32 is leaving his tent, carrying a plastic woman in his arms. According to the caption, the man is "Saving the most precious first" (Catalogue, 2007, p. 36).

The Monument to Waiting Woman in Murmansk city is just one example of the role reserved for women in the Arctic. This woman made of stone is looking to the sea as if waiting for her husband to come back home; it is a telling example of how the memory of the Arctic is gendered. From the point of view of Arctic history, the photographic exhibition, too, was a logical continuation of the image that the Russian Arctic used to have and in part still has.

Understanding the Russian Arctic as an identity project?

The Arctic megaproject can be interpreted as purely a development enterprise of a strategic region which is about to provide Russia with natural resources and economic independence. The flip side is a region-building process for the purposes of identity politics. It is a project that entails questions of social and cultural sustainabilities from various angles, which are still to be resolved if Russia wants to aim at sustainable development in the Arctic.

With a decreasing population in most of the Arctic regions, a major question in the megaproject has been the availability of the workforce and the social sustainability of Arctic cities (Heleniak, 2017). The Russian government has responded to the loss of especially young people from the Far North by establishing a program within Rosmolodëž', a state-driven organization for promoting national patriotism for younger generations in order to raise them to be loyal citizens and future decision-makers in the state administration (Rosmolodëž', no date). Different youth programs are intended to educate loyal citizens whose identity partly draws on the idea of the Great Russian Arctic and a certain arctic identity. These endeavors show that Russia needs more (young) people to engage their lives with the Arctic. Historical culture reveals that the Arctic is much more to Russia and Russians than a massive exercise to rule "the immense emptiness," which covers almost one sixth of the world's land surface.

In the light of Russian identity politics, it is no wonder that everyday life does not serve as a source for national (imperial) identity. Instead, celebrating the boundless space that only Russians (men) have the ability to conquer and stressing the crucial role of the North in the future of Russia from economic and military perspectives seem to be the priorities in Russian Arctic politics.

In the process of imagining the vast Russian Arctic, historical resources support its position as a shelter and guarantor of social order, stability and wealth, which are important elements of both national identity and national security. Some scholars (Karaganov et al., 2014, p. 62) have called the development of the Arctic not a mega-, but metaproject that aims at the historical mission of Russia to serve as a bridge between Europe and Asia.

By keeping the Arctic at a distance, as utopias usually are distant, the Russian political elites replicate the "strongest and most beautiful Russian myth about taming the elusive frozen lands" (Karaganov et al., 2014, p. 62). Russian historical culture of the Arctic replicates this myth by presenting and reproducing the Arctic as an unambiguously heroic arena. The historical culture reveals to us an integral part of national Arctic politics and strengthens the official narrative. In this sense, it enhances culturally sustainable development and legitimizes the chosen political measures – and ought to be taken into consideration when analyzing Russian Arctic politics.

References

Ahonen, S. (1998) *Historiaton sukupolvi? Historian vastaanotto ja historiallisen identiteetin rakentuminen 1990-luvun nuorison keskuudessa*. Helsinki: Suomen Historiallinen Seura.
Baev, P.K. (2013) "Russia's Arctic ambitions and anxieties," *Current History*, 112(756) [Online]. Available at: www.currenthistory.com/pdf_org_files/112_756_265.pdf (Accessed: April 24, 2019).

Barents Observer (2017) *Take a look inside Russia's northernmost Arctic military base*, April 18 [Online]. Available at: https://thebarentsobserver.com/en/security/2017/04/take-look-inside-russias-northernmost-arctic-military-base (Accessed: April 24, 2019).

Barents Observer (2018) *Novaâ goskomissiâ pristupaet k ispolneniju bolshogo plana Putina po Arktike*, October 9 [Online]. Available at: https://thebarentsobserver.com/ru/arktika/2018/10/novaya-goskomissiya-pristupaet-k-ispolneniyu-bolshogo-plana-putina-po-arktike (Accessed: April 24, 2019).

Borén, T. (2009) *Meeting-places of transformation: urban identity, spatial representations and local politics in post-Soviet St Petersburg*. Stuttgart: Ibidem.

Catalogue of the exhibition *Arctic!!!* (2007) Moscow: Moscow House of Photography.

Dahl, J. (2012) "The constitution and mobilisation of political power through utopian narratives in the Arctic," *The Polar Journal*, 2(2), pp. 256–273.

Eng, T. and Lindaräng, I. (2013) "Negotiating local, national and Nordic identities through commemorations" in Aronsson, P. and Gradén, L. (eds.) *Performing Nordic heritage: everyday practices and institutional culture*. London and New York: Routledge, pp. 99–128.

Fedorov, V.P. (2018) *Rossiâ: arktičeskij resurs* [Online]. Available at: www.instituteofeurope.ru/images/uploads/analitika/an97.pdf (Accessed: April 24, 2019).

Forest, B. and Johnson, J. (2002) "Unraveling the threads of history: Soviet-era monuments and post-Soviet national identity in Moscow," *Annals of the Association of American Geographers*, 92(3), pp. 524–547.

Heleniak, T. (2017) "Boom and bust: population change in Russian Arctic cities" in Orttung, W.R. (ed.) *Sustaining Russia's Arctic cities*. New York and Oxford: Berghahn, pp. 67–87.

Horensma, P. (1991) *The Soviet Arctic*. London: Routledge.

Karaganov, S. et al. (2014) *Nacionalnaja identičnost' i budušee Rossii. Doklad Meždunarodnogo diskussionnogo kluba "Valdaj"* [Online]. Available at: http://vid1.rian.ru/ig/valdai/doklad_identichnost_RUS_ISBN.pdf (Accessed: April 24, 2019).

Komitet (2018) *Komitet Sankt – Peterburga po delam Arktiki* [Online]. Available at: www.gov.spb.ru/gov/otrasl/arkt (Accessed: April 24, 2019).

Komsomolskaâ Pravda (2017) *Na Vasil'evskom ostrove otkryt pervyj v Rossii pamâtnik polârnikam*, September 15 [Online]. Available at: www.spb.kp.ru/daily/26732.7/3758684/ (Accessed: April 24, 2019).

Laruelle, M. (2014) *Russia's Arctic strategies and the future of the far North*. Armonk, NY: M.E. Sharpe.

Lempinen, H. and Heininen, L. (2017) "Paikallisten elämäntyylit, alkuperäiskansojen kulttuurit? Kulttuuri ja sen kestävyydet arktisten valtioiden strategioissa," *Alue ja ympäristö*, 45(1), pp. 4–14 [Online]. Available at: https://aluejaymparisto.journal.fi/article/view/60678 (Accessed: April 24, 2019).

Lempinen, H. (2018) "Ovatko kulttuurit vain välineitä arktisille valtioille?" *Versus*, May 22 [Online]. Available at: www.versuslehti.fi/tiededebatti/ovatko-kulttuurit-vain-valineita-arktisille-valtioille/ (Accessed: April 24, 2019).

Medvedev, S. (2018) "Simulating sovereignty: the role of the Arctic in constructing Russian post-imperial identity" in Tynkkynen, V-P., Tabata, S., Gritsenko, D. and Goto, M. (eds.) *Russia's far North: the contested energy Frontier*. London and New York: Routledge, pp. 206–215.

Metodičeskie rekomendacii po organizacii i provedeniû v obšeobrazovatelnyh organizaciâh Rossijskoj Federatsii Vserossijskogo uroka "Arktika – Fasad Rossii" (2016) Moskva.

Mjør, K.J. (2017) "Nikolai Berdiaev and the "boundless spaces" of Russia," *Nordlit*, 39, pp. 4–17.

MK.ru (2007) Artur Čilingarov: "*My dokazali – Arktika naša*" [Online]. Available at: www. mk.ru/editions/daily/article/2007/08/07/88555-artur-chilingarov-myi-dokazali-arktika-nasha.html (Accessed: April 24, 2019).

Ned Lebow, T. (2008) "The future of memory," *The ANNALS of the American Academy of Political and Social Science*, 617, pp. 25–41.

Nyyssönen, H. and Halla, M. (2013) "Muisti, kokemus ja historiallinen käänne kansainvälisessä politiikassa: näkökulmana Richard Ned Lebow ja Reinhart Koselleck," *Kosmopolis*, 43(2), pp. 21–38.

Paasi, A. (2002) "Bounded spaces in the mobile world: deconstructing 'regional identity'," *Tijdschrift voor Economische en Sociale Geographie*, 93(2), pp. 137–148.

Pirnes, S. (2017) "Pelkoa ja uhkakuvia. Suomalaisen median kuva Venäjän toiminnasta arktisella alueella," *Lähde—historiatieteellinen aikakauskirja*, pp. 63–81.

Riukulehto, S. (2015) "Mitä on aluehistoria?" *Maaseudun uusi aika*, 3, pp. 39–53.

Rogachevski, A. and Steinholt, Y. (2017) "'Boundless' Russia and what to make of it," *Nordlit*, 39, pp. 1–3.

Rosmolodëž' (no date). *Official website*. Available at: https://fadm.gov.ru/.

Šabaev, Û.P., Sadohin, A.P. and Kuznetcova, A.Û. (2016) "Rossiskaâ identičnost Russkogo severa: istoriâ i problemnaâ sovremennost," *Vestnik SPbGU*, 12(1), pp. 127–140.

Sergunin, A. and Konyshev, V. (2014) "Russia in search of its Arctic strategy: between hard and soft power?" *The Polar Journal*, 4(1), pp. 69–87.

Stammler, F. (2019) Personal communication.

Tilli, J. (2009) "Tiloja, linjauksia, retoriikkaa—historiapolitiikan ulottuvuuksia," *Historiallinen aikakauskirja*, 3, pp. 280–287.

Timtchenko, L. (1997) "The Russian Arctic sectoral concept: past and present," *Arctic*, 50(1), pp. 29–35.

TripAdvisor, www.tripadvisor.ru/Attraction_Review-g298501-d8464961-Reviews-Monument_to_The_Frontier_Guards_of_Arctic-Murmansk_Murmansk_Oblast_Northwestern_D.html (Accessed: April 25, 2019).

Vodolazkin, E. (2018) *The aviator*. London: Oneworld Publications.

Widdis, E. (2003) *Visions of a new land: Soviet film from the revolution to the Second World War*. New Haven and London: Yale University Press.

Yakhina, G. (2019) *Zuleikha*. London: Oneworld Publications.

13 The resourceful North

Divergent imaginaries from the European Arctic

Monica Tennberg, Hanna Lempinen
and Susanna Pirnes

As Gail Fondahl and Gary Wilson (2017, p. 4) have pointed out, "northern regions and communities have their unique histories and problems – a reality that has set them on different trajectories in terms of development. There are numerous sustainabilities and numerous norths." This reality is seldom reflected in the scholarly literature or popularized debates, which tend to depict the Arctic region as one, not many. The notions of sustainability and sustainable development are typically treated in equally elusive and ubiquitous ways, ignoring both the conceptual complexity and the practical challenges that "successfully" applying these notions in practice entail. In this volume, we have – in contrast with the mainstream discourses on sustainability in the Arctic, which focus heavily on natural resources – taken a more rounded approach to addressing and acknowledging the multiple dimensions of social and cultural sustainabilities and the ways in which they are entangled with material and human resources alike. Geographically, the chapters of this book have shed light on the diversity of environmental, social and cultural sustainabilities in a range of case studies from the broadly defined European Arctic, in particular from Finland, Greenland and Russia. The different kinds of data from a variety of sources – mainly analysis of documents, interviews and textual and visual media material – and the combining of different research methods – mostly based on qualitative, relational and socio-material approaches – have sought to provide us more insights into and a better understanding of social and cultural sustainabilities.

While the chapters of this book have dealt with a wide scope of social and cultural sustainability concerns ranging from natural resource and climate policies to regional and human development, knowledge production, cultural appropriation and beyond, they have something in common: they all make observations about the ways in which the issues of social, cultural and development are entangled with the broader thematics of self-determination and national pride and the understanding of good life and happiness. These observations work to highlight not only the diversity and multidimensionality of "real-world" sustainability concerns, but also the artificial nature of slicing sustainability into the separate dimensions of the social, cultural, economic and environmental. Indeed, it can be argued that no such thing as a social or cultural dimension can be separated and neatly cut from the overall sustainability concern. Sustainability, despite its techno-scientific legacies and connotations (see e.g. Newberry, 2013), is ultimately always a social concern as it has to do with *continuity* – the very question of maintaining life as we know it and securing things that we hold dear (Kassel, 2012; Missimer et al., 2010; Sorsa, 2015).

The ways that social and cultural sustainability concerns entwine and entangle with resources of different kinds and broader issues of regional development place them in a central position in the European Arctic as a popular object of social imaginaries for future material prosperity and human wealth and well-being (cf. Appadurai, 1996; Taylor, 2004; Jasanoff, 2015; Vehkalahti, 2017; Kristoffersen and Langhelle, 2017). Such imaginaries contribute to the

> ways people imagine their social existence, how they fit together with others, how things go on between them and their fellows, the expectations that are normally met, and the deeper normative notions and images that underlie these expectations.
>
> (Taylor, 2004, p. 23)

Although they often appear impenetrable, such social imaginaries are, in fact, partial, biased and incomplete and often sources of tensions and disruptions. All the case studies on northern sustainability featured in this book also shed light on these complexities as constraining or enabling social imaginaries of sustainability or both.

Resource-intensive imaginaries

In the resource-intensive social imaginaries associated with materialities (say, energy and mineral resources), technologies (such as nuclear and hydropower), the imaginaries become embedded and entwined in our culturally laden understandings and idea(l)s of the good life and desirable futures, communities and societies. The contributions in this part of the volume discuss energy and mineral resources as important sources of income, employment and economic development in the Arctic but also as a question of security, nationalism and climate politics, providing examples of powerful, attractive social imaginaries for social sustainability.

Marjo Lindroth's chapter demonstrates the power of effects that are entailed in envisioning resources and how they might contribute to a better future in Greenland. The hopes for large-scale resource extraction enable imaginaries of independence, wealth and self-sufficiency. Such strong visions render the economic aspect of the country's future development "rational" and legitimate. The futures and resources that are excluded through the economically focused imaginaries have to do with Greenland's cultural, environmental and social questions. Accordingly, these aspects and their sustainability become sidetracked, and other imaginaries – ones that fall outside the economic focus – often come to be labeled as "irrational," even unpatriotic, in the resource visions of the country.

The (future) construction of a nuclear power plant in Pyhäjoki, as Hannah Strauss-Mazzullo shows, is enabled by a community willing to take on the burden of living with the risk of nuclear accidents. Further, it is enabled by government support and the need to cut greenhouse gas emissions, and it is enabled by a technological know-how and long-term safety record Finland can rely on. Hence, a nuclear power megaproject is important far beyond local development. The size of the project means that not just the community but the whole country gets locked into the development of the nuclear path, in contrast to other possible solutions. The project also crosses the national border in the form of electricity, with repercussions in those countries participating in the Nordic electricity market and beyond. While it means that Finland will produce less greenhouse gas emissions by making redundant the burning of fossil fuels for electricity production once the nuclear power plant is up and running, it also means that the project constrains alternative energy solutions.

The chapter by Joonas Vola concludes that, in his case study, the Kemi River spoken of as a resource is an attempt of ordering the river and the inhabitants along it through a process of rationing and reasoning by language. The contingent nature of the river is challenged as unsustainable for the social order and cohesion of the community. The river needs to be turned into a potentiality to make it communicative and orderly to manage another resource: that is, the human resource. The language follows a grammar and sentences linearity, starting with a capital letter and ending with a period, restraining the social imagination in between. The river as a resource, indicating potentiality, actualizes at the cost of alternative imaginaries for the river and the people along it. To alter an image, one must speak in terms of "may" instead of "will."

In her closely related chapter, Hanna Lempinen explores how social sustainability becomes entangled with the today and tomorrow of the Arctic energy concern with a specific focus on the often-sidelined aspect of energy as a social sustainability concern in the circumpolar North. In her analysis, the framing of the regional energy concern both enables and constrains the ways in which the region and its future development prospects are thought of. On the one hand, the regional energy discourse makes possible a gradual transition toward less carbon-intensive and more climate-friendly Arctic energy production; on the other hand, the same discourse continues to frame the region's future prospects in terms of its energy industries and activities, leaving little to no room for imagining alternative pathways to regional sustainability and development.

Adrian Braun suggests in his chapter that municipalities, cities and companies in the European Arctic have diverse possibilities to strive for social and ecological sustainability. Green financial products, such as socially responsible investments, promise to be a capital-raising tool to improve the quality of lives and ecological footprints. This makes room for plenty of social imaginaries for a myriad of actors involved. Green capital, as raised, for example, via climate bonds, may not only enable communities to develop schemes for green housing, energy-efficient transport and other eco-friendly solutions, but also enable visions for the global investment community to generate double benefit, comprising profits and reputation building. The social imaginations in the European Arctic feed a wealth of green project ideas, and there is a need to overcome the imaginary constraints which make the global finance sector "still" overlook the Arctic. Global investors do not wish to neglect opportunities. Thus, the combination of green investments and current challenges in the Arctic enables new opportunities in social, ecological, economic and cultural perspectives.

Paula Tulppo discusses in her chapter the mismatch between European and local imaginaries on cross-border cooperation and the necessary resources for its development. While the EU's cross-border cooperation program Interreg V A Nord focuses on economic resources and employment as a basis for regional development, the municipalities appreciate resources that promote the good everyday life, based on a combination of economic, social and cultural factors. The current focus on economic resources, while recognized as important for regional development, also locally constrains the imagination for sustainability. The analysis suggests that cultural diversity aids regional cohesion and concludes by asking how to maintain the region's cross-border culture, which stems from a shared history and cooperation.

Whose imaginaries?

Knowledge in different forms fuels social imaginaries. Sheila Jasanoff (2004, p. 3) emphasizes the importance of politics in co-production: that is, "how knowledge-making is incorporated into practices of state-making, or of governance more broadly, and, in reverse, how practices of governance influence the making and use of knowledge." As the Arctic is being imagined and mediated by a wide array of actors, it is necessary to study the role of different knowledges in social imaginaries created by different actors on various platforms if we are to understand their diversity and identify any disruptions and tensions.

Gemma Holt in her contribution points out that the politics of adaptation and knowledge are closely linked by questions of *whose* knowledge is relevant for adaptation and how that knowledge can be effectively communicated and linked to policy and practice. In her analysis of a report on adaptation actions for the Barents region, she notes that non-Western knowledge, while recognized as important and relevant for adaptation, remains "the other." This has bearing on the sustainability of traditional forms of knowledge; while they will continue to be embedded in specific cultures, they may never be integrated into adaptation strategies. Adaptation decisions that are made without integrating multiple knowledges have a direct effect on the construction of social imaginaries of sustainability. Therefore, the politics of both knowledge and adaptation must be transformed in order to increase long-term sustainability for the people and the environment of the Barents region and beyond.

The chapter by Monica Tennberg investigates the social imaginaries of economic development and growth embedded in recent national and regional discourses about Arctic expertise and smart specialization in Lapland. These popular discourses celebrate the skills, creativity and innovations of Northern inhabitants, but the crucial social dimensions of Arctic expertise and smartness are largely dismissed in the Finnish programs and plans. The issue, in short, is whether such expertise and smartness can be developed and sustained without experts themselves. Lower levels of education, lack of investments in research and development and poorly developed innovation environments, among other things, hamper the fulfillment of social imaginaries that the Arctic expertise and smartness in Lapland builds upon.

Heidi Sinevaara-Niskanen illustrates how questions of gender fall under the radar of the contemporary imaginaries of Arctic resources. As the chapter points out, for the past twenty years there has been a growing scientific and political awareness of the role that gender plays in development in the Arctic. Feminist scholarship in particular has aimed at demonstrating how gender equality as well as recognizing gendered knowledge and intersectionality could support the sustainability of the region. This knowledge of issues of gender and the newly emerged focus on social resources has so far not significantly changed the perceptions of the region's resources. Questions of gender are yet to make their way to the predominant imaginaries of what constitutes resources in the Arctic.

In his contribution, Francis Joy discusses the cultural appropriation of Sámi cultural heritage, religious traditions and practices in the Finnish tourism industry. Some of the most sacred aspects of Sámi culture, such as drums, which are intimately linked with the practice of shamanism, as well as traditional costumes, are exploited to serve the needs of the tourism industry in Lapland. An illusion of sustainability thus emerges through the misuse of Sámi traditions for tourism development. The active marketing efforts pertaining to the cultural heritage of the Sámi people constrain Sámi traditions and contribute to the assimilation of their culture and history into the Finnish mainstream.

Susanna Pirnes discusses how the use of history enables the Russian Federation to build up Arctic identity. Historical culture offers the site through which people meet their past and imagine the Arctic. The idea of limitless space provides resources for imagination of power, which ties up with the aim of the political elite to present the Arctic as an integral part of Russian identity for the purposes of the legitimation of politics. The elite steers the development into one (political) Arctic and not for many (local) Arctics. Uses of Arctic history unfold in various ways: areal resources support the image of the Arctic as a shelter and guarantor of social order, stability and wealth. These are important parts of national identity but also issues of national security and pride.

Final remarks

This book was written to raise the importance of social and cultural sustainability and its different aspects both in the context of Arctic research and in terms of discussions about the Arctic in general. We have sought to diversify the ways in which "resources" in the context of the region are understood: the European Arctic is not only rich in resources, but also *resourceful* in terms of its social and cultural resources and their potentialities. Thus, instead of treating sustainability as a political concept, embedded in the national and international discursive practices of states, non-governmental organizations and other institutions (Pram Gad et al.,

2019), we propose to treat (social) sustainability as a practice beyond politics: as a social practice and a way to understand the world. Practices of social imagination both enable and constrain our understanding of what is preferred and possible, but the analysis of such practices also draws our attention to our histories, power relations and epistemologies closely connected to our imagination for sustainability.

A social imaginary is that "common understanding that makes possible common practices and a widely shared sense of legitimacy" (Taylor, 2004, p. 24). Our social imaginary at any given time is complex, both factual and "normative." According to Taylor (2004, p. 24), "we have a sense of how things usually go, but this is interwoven with an idea of how they ought to go." As "our" shared social imaginaries about the Arctic are being created now, we need to understand what kind of imaginaries are popular and attractive and who gets to be a part of them, not least because the imaginaries are shared and debated extensively by the political actors and various stakeholders inside and outside the region. Indeed, these imaginaries and this world are not without conflicts and disagreements as our contributions show – between values, ideals, different agencies and institutions. Currently, the internationally shared imaginaries of the Arctic as the home of the polar bear or as the uninhabited and infinite pool of natural resources fail to resonate with the (g)locally lived and experienced realities in the European Arctic. From the perspective of the region's residents themselves, such imaginaries are often limited, narrow and misrepresentative in terms of the local diversity of identities, lives, experiences and sustainability concerns. Through a critical discussion, our book has aimed to contribute to and diversify these narrow and simplistic "image-based, narrative-based accounts of strips of reality" (Appadurai, 1996, p. 35) that continue to construct and convey the popular imaginaries of the region. In addition, what potentially and hopefully unfolds from the different social imaginaries discussed in the book are ideas and opportunities that lead to more sustainable regional futures and local actions.

References

Appadurai, A. (1996) *Modernity at large. Cultural dimensions of globalization.* Minneapolis and London: University of Minnesota Press.

Fondahl, G. and Wilson, G. (2017) "Exploring sustainabilities in the Circumpolar North" in Fondahl, G. and Wilson, G. (eds.) *Northern sustainabilities: understanding and addressing change in the circumpolar world.* Cham: Springer, pp. 1–9.

Jasanoff, S. (2004) "The idiom of co-production" in Jasanoff, S. (ed.) *States of knowledge: the co-production of science and the social order.* London: Taylor and Francis, pp. 1–12.

Jasanoff, S. (2015) "Future imperfect: science, technology, and the imaginations of modernity" in Jasanoff, S. and Kim, S-H. (eds.) *Dreamscapes of modernity: imaginaries and the fabrication of power.* Chicago: University of Chicago Press, pp. 1–33.

Kassel, K. (2012) "The circle of inclusion: sustainability, CSR and the values that drive them," *Journal of Human Values*, 18(2), pp. 133–146.

Kristoffersen, N. and Langhelle, O. (2017) "Sustainable development as a global-Arctic matter: imaginaries and controversies" in Keil, K. and Knecht, S. (eds.) *Governing Arctic change. Global perspectives.* London: Palgrave MacMillan, pp. 21–41.

Missimer, M., Robèrt, K., Broman, G. and Sverdrup, H. (2010) "Exploring the possibility of a systematic and generic approach to social sustainability," *Journal of Cleaner Production*, 18, pp. 1107–1112.

Newberry, D. (2013) "Energy affects: proximity and distance in the production of expert knowledge about biofuel sustainability" in Strauss, S., Rupp, S. and Love, T. (eds.) *Cultures of energy: power, practices and technologies*. Walnut Creek: Left Coast Press, pp. 227–241.

Pram Gad, U., Jakobsen, U. and Strandsbjerg, J. (2019) "Introduction: sustainability as a political concept in the Arctic" in Pram Gad, U. and Strandsbjerg, J. (eds.) *The politics of sustainability: Reconfiguring identity, space, and time*. London: Routledge, pp. 13–23.

Sorsa, V-P. (2015) "Johdatus symposiumiin: kestävän talouden ja vahvan kestävyyden jäljillä," *Poliittinen talous*, 3(1), pp. 11–28.

Taylor, C. (2004) *Modern social imaginaries*. London: Duke University Press.

Vehkalahti, P. (2017) *Pohjoisen ydinmylly: julkinen keskustelu Fennovoiman ydinvoimalasta 2007–2013*. Tampere: Tampere University Press.

Index

accountability of policymakers 107
Act on Greenland Self-Government (2009)
    16–17, 19
activists 28
actualization of Greenland's
    independence 22
adaptation 103; and policy-relevant
    knowledge 108; politics of 108, 178;
    and sustainability 103–104
Adaptation Actions for a Changing Arctic
    (AACA) 105, 107, 114–115; analyzing
    the treatment of multiple knowledges
    in 109–112; authors of 106; Barents
    report 105–106; implications of
    111–112; saliency of 111; summary for
    policymakers 111–112
affective elements 6, 23
"affective resonances" 16
affordability of energy resources in the
    Arctic 59–60, 63
Agamben, Giorgio 49
Ahmed, S. 36
Akademik Lomonosov 33
Alt, Suvi 44, 52
analyzing the treatment of multiple
    knowledges in the AACA 109–112
appropriation of Sámi culture: contributing
    factors 147–148, 150; Sámi voices
    speaking out against 151, 152–153
Arctic, the 9; adaptation strategies for
    106; affordability of energy resources
    in 59–60, 63; availability of energy
    resources in 59–60, 63; border
    regions 84; challenges of climate
    change 63; cultural sustainability 164;
    curbing greenhouse gas emissions
    64; development of 83; discourse on
    resources and gender 137–138; energy
    reserves vs. regional energy security

needs 58–60; energy resources 7, 56;
    extractive energy development 60;
    gendered memory of 171; historical
    knowledge 9; homogeneity of
    communities in 3; implementation
    of energy development projects
    61; Kemijoki River 6–7; natural
    resources 2, 4, 8; Northern Sea Route
    Administration 165; politics of 131;
    prudent development of resources in
    61–63; radioactive contamination of 34;
    regional energy development 59–60, 61;
    Russian Arctic Zone 163–164; Russian
    claims to the continental shelf 165;
    Russian military activity in 165; Russian
    North 167; scientific exploration of 104,
    171; social resources 131–132; social
    sustainability 164; socially responsible
    investments (SRIs) 7; as space
    167–168; sustainability 5; sustainable
    development of it's energy resources 60;
    technological development of 59–60
Arctic Climate Impact Assessment
    106, 135
Arctic Commission of St. Petersburg
    164, 165
Arctic Council 8, 57, 63, 104, 105, 109,
    131, 133, 134, 139; Adaptation Actions
    for a Changing Arctic (AACA) 8;
    Arctic human development report 133;
    Arctic Monitoring and Assessment
    Programme (AMAP) 105; involvement
    of indigenous peoples 107; legitimacy
    of 110–111; member states 104;
    Sustainability Working Group 58;
    Sustainable Development Working
    Group 134
Arctic Energy Summits 57, 58, 59–60;
    final reports 58, 59–61, 62, 63, 64

Arctic Environmental Protection Strategy 104
Arctic expertise 8, 117–118, 119, 123, 127, 128; in Finland 119
Arctic human development report 133, 135; feminist analysis of 134
Arctic Monitoring and Assessment Programme (AMAP) 105; summary reports 106
Arctic Renewable Energy Atlas (AREA) 59
Arctic research and policy reports 135
Arctic studies 2
Arctic sustainability science 2, 6
Association of Polar Explorers 169–170
Atropos 49, 50
availability of energy resources in the Arctic 59–60, 63

Barents Area Overview Report 110–112
Barents region 105; radioactive material in 34
Barents report 105–106
Berdyaev, Nikolai 163
Berit Dahlbakk, Ellen 156
bios 49
Bjørst, L. R. 21
bohemian index 124
bonds 75; and climate bonds 75–76
border regions in the Arctic 84
"bow" 48–49
"bow of Heraclitus" 48
"bow of life" 48
Braun, Adrian 7, 177
Brundtlan Commission 61
Bydler, C. 147

capital 70; creative 123–124; green 72, 177; human 72, 123; social 123
capitalism 47; "self-destruction thesis" 47
carbon footprints 69
Carney, Mark, "Breaking the tragedy of the horizons: climate change and financial stability" 74
Cetina, Knorr 35
China: climate bonds 80; Green Financial System 74; issuance of climate bonds 76
Čilingarov, Artur 165, 169
Čkalov, Valery 168
clean energy 71
climate bonds 74–76; issuance 75; issuance of in Denmark 77–78; issuance of in Finland 78; issuance of in Iceland 76; issuance of in Norway 76–77; issuance

of in Russia 78–79; issuance of in Sweden 77; versus regular bonds 75–76
Climate Bonds Initiative 76, 77
climate change 7, 21, 37, 56, 63, 64, 69, 74, 106; adaptation strategies 103; Arctic climate impact assessment 135; in the European Arctic 74; final reports from the Arctic Energy Summits on 64; and flood risk prevention 46; and greenhouse gas emissions 64; mitigation strategies 103; policymakers 107; socially responsible investments (SRIs) 7; *see also* adaptation
climate policies 7
Clotho 49
coercion 45
colonialism 150
Colpaert, A. 34
commodities 46
communities 22–23; cultural sustainability 4; nuclear power production in 37; Pyhäjoki 28; social sustainability 3–4; as stakeholders in Arctic development 62; support for high-risk technology in 32; *see also* Pyhäjoki
Conference of Parties (COP) 74
content analysis 91
contingency 44; of the Kemi River 47; pure 45
corporate social responsibility (CSR) 69, 70; in the European Arctic 72
cors 45
course 45
creative capital 123–124
creativity in Lapland 123–125
credibility, of AACA policy 110
cross-border programs 87, 88, 96
cultural appropriation 146; consequences of using Sámi cultural heritage as a resource 157, 159; of *Duodji* 147; misuse of Sámi culture for tourism 155–157; of Sámi heritage 147–148, 150; and the Sámi struggle for recognition 155; Sámi voices speaking out against 152–153
cultural development 97
cultural resources, of the Interreg VA Nord program region 93–94
cultural sensitivity 160
cultural sustainability 2, 3, 5–6, 7, 9, 164, 166, 175, 176, 179–180; divination drums of the Sámi people 144–145; *Duodji* 145
culture 4, 5

Curie, Marie 35
cursus 45

day trading 74
de Jong, C. 146
Dean, Mitchell 119
demonstrations 32, 38
Denmark 6, 15; Act on Greenland
    Self-Government (2009) 16–17,
    19; actualization of Greenland's
    independence from 22; annual block
    grant to Greenland 19; Greenland's
    pursuit of independence from 16–18,
    19–20, 21, 22; issuance of climate
    bonds in 77–78; power relations with
    Greenland 23; resource extraction in 22;
    uranium mining 22
development 2, 7, 8, 24, 82; actors 83;
    and "affective resonances" 16; cultural
    97; endogenous 86, 87; exogenous 86,
    97; in Finland 33; of Finnish Lapland
    33; human 8; Mankala principle 31;
    of nuclear power 28; opposition to
    resource projects 23–24; regional 86,
    95, 97; of Rovaniemi 51; of Salla 3–4;
    siting procedure for nuclear power
    plants 32–33; socially unsustainable 3;
    technological 59–60; *see also* regional
    development; sustainable development
disagreements 3–4
discourse 45, 49, 50
discourse analysis 41
disposal of nuclear waste 34, 37
divination drums 144; consequences
    of using Sámi cultural heritage as
    a resource 157, 159; faked 145;
    reproduction of fakes in Finland
    152–153; symbolism of 144–145, 151
Dodds, K. 18
domestic production of uranium 36
Domínguez, L. 87
Duodji 145; cultural appropriation of 147;
    and misuse of Sámi culture for tourism
    155–157

ecological knowledge 106, 107
economics: of extracting Greenland's
    resources 20; opposition to resource
    projects 23–24; of resource
    extraction 23
ecosystem services 69
endogenous development 86
endogenous regional development 87

energy resources 7; in the Arctic 56; Arctic
    Energy Summits 57–58; and fossil fuel
    emissions 63; as regional concern 57
Enontekiö 84; development strategies 91;
    see also Interreg V A Nord program
entrepreneurship index 124
entwinement of resource extraction and
    pursuit of independence in Greenland 23
equality 8, 132; first-wave feminism 136;
    gender 135–136, 137–138; income 120;
    and intersectionality 139–140; second
    wave feminism 136; Taking Wing
    conference 134; third wave feminism
    136–137
ESG 70, 73, 74
esourse 42
ethnicity 132; and gender 139
etymology: of "course" 45; of "economy"
    47; of "Fates" 50; of "resource" 42
Eurajoki 6, 32, 35, 37
European Arctic 7, 70, 177–178;
    challenges faced by 74; climate bond
    issuance 75; Denmark, issuance of
    climate bonds in 77–78; extractive
    industries 72; Finland, issuance of
    climate bonds in 78; Iceland, issuance
    of climate bonds in 76; natural resources
    72; need for socially responsible
    investments 71; Norway, issuance
    of climate bonds in 76–77; Russia,
    issuance of climate bonds in 78–79;
    socially responsible investments (SRIs)
    in 79; Sweden, issuance of climate
    bonds in 77
European Commission 31
European Union (EU) 7, 22, 36, 86, 127;
    cross-border program 84; cross-border
    programs 88, 96; directives 46, 105;
    flood risk prevention expenditures
    44, 46; Interreg A 87; Interreg V
    A Nord program 83–84, 88; Regional
    Social Progress Index 120, 121; water
    management 45–46; *see also* Interreg V
    A Nord program
exclusion of women in polar research
    activities 133–134
exogenous development 86, 97
expertise: Arctic 8, 117–118; regional 110,
    119; social dimensions of 119–120;
    see also Arctic expertise

fake divination drums 9, 145
fari 50

fata 50
"Fates" 49, 50
feminist theory 132; first-wave feminism
136; gender equality 132–133;
intersectionality 136–137, 139–140;
recognizing particular vulnerabilities
138–139; third wave feminism
136–137
Fennovoima company 30, 31, 32–33
field experiments 35
final reports from the 2007–2017 Arctic
Energy Summits 57, 58, 59, 65; on
climate change 64; on sustainable
development 60–63
financial products: day trading 74; ESG
performances 76; "green" 74–76, 79
Finland 7; aging population of the
research border region 84; Agreement
of Friendship, Cooperation and Mutual
Assistance 30; Arctic expertise 119;
Arctic strategy 120; energy policy 29;
Enontekiö 84; Eurajoki 34, 35; fake
divination drums as marketing tool for
tourism 145; Fennovoima company
30–31; flood risk prevention 46; forestry
sector 29; hydropower plants 44;
institutional cross-border cooperation
88; Interreg V A Nord program 88;
issuance of climate bonds in 78; Kemi
River 43, 47, 49; Kemijärvi 33; Kolari
84; Local Government Act 90; local
governments 46; Mankala principle
31; misuse of Sámi culture for tourism
155–157; municipalities 84, 88, 90,
96; Muonio 84; nuclear history 30, 31;
nuclear power production 29; nuclear
waste management 34; Olkiluoto 3 31;
Ounasjoki 43; Paukkajanvaara 34–35;
Pello 84; privatization of flood insurance
47; Pyhäjoki 28, 37–38; Radiation and
Nuclear Safety Authority (STUK) 31;
Raudanjoki 43; regional development
policy 86; reproduction of fake Sámi
drums in 151, 152–153; resources 29;
risks of living with nuclear power plants
27; Rovaniemi 43, 52; Sámi 9; Simo
33; siting procedure for nuclear power
plants 32–33; support for high-risk
technology in 32; Tornio 84; Tornio
River 46; unemployment rates of the
research border region 85; uranium as
domestic product 36; uranium mining
28; Visit Arctic Europe 153; Ylitornio

84; *see also* Interreg V A Nord program;
Sámi people
Finnish Lapland 41; development of 33;
industrial investments 79; Kemijoki
River 6–7; social sustainability in 118
Finnish-Swedish Arctic border 84
first-wave feminism 136
flood insurance, Finland's privatization
of 47
flood management 42; on the Kemi
River 50
Floods Directive 45
Fohndahl, Gail 175
Fonneland, Trude 157
Fortum 34
fossil fuel extraction 63–64
Foucault, Michel 41; governmentality 119
fuel cycle 34
Fukushima 28, 33

Gad, Ulrik Pram 23
Gákti 147
Gákti costume 9
gender 8, 132, 140, 179; and Arctic
politics 132; exclusion of women in
polar research activities 133–134; first-
wave feminism 136; intersectionality
136–137, 139–140; recognizing
particular vulnerabilities 138–139; as
resource 134; and resource management
in the Arctic 137–138; second wave
feminism 136; Taking Wing conference
134; third wave feminism 136–137;
women's representation in politics
132–133; *see also* feminist theory
genealogy of language 41
GINI co-efficient 120
Gjørv, Hoogensen 132, 134, 135
global energy 56
Gorbačëv, Mikhail 104
Gosudarstvennyj sovet 165
governmentality 119
Great Patriotic War 166, 170
"green" financial products 74–76
"green" investments 7, 73; performance
of 73
green investments 71, 72
greenhouse gas emissions 64, 69, 71,
103, 177
Greenland 6, 16; attracting investment in
19–20; block grant from Denmark 19;
and climate change 21–22; communities
22–23; development of 24; entwinement

of resource extraction and pursuit
of independence 16; experiences of
inequality 22–24; exploration licenses
20, 21; extraction licenses 20; goal
of independence 15, 16–18, 19–20,
21, 22, 24; Inatsisartut 15; industrial
investments 79; Inuit population 21;
natural resources 19; oil and mineral
strategy 18, 19; opposition to resource
projects 23–24; politics of development
in 20; power relations with Denmark 23;
resource extraction in 17, 22; resource
spaces 19–20; resources 18; uranium
mining 23; zero tolerance policy on
uranium mining 15
gross regional product (GRP) 120
Grzanka, P. R. 155
Guerrieri, Valeria 22
Gulag labor camps 168
Guttorm, Gunvor 147, 156

Habeck, J. O. 133
Hamina Peace Treaty 84
happiness 36–37
"happy object," uranium as 36
Harlin, Eeva- Kristiina 159
Hassi, Satu 31
Haugaard, Mark 45
Hautala, Heidi 31
Helander-Renvall, E. 151
Heraclitus 48
high-risk technology 29, 34; disposal of
nuclear waste 37; domestic production
of uranium 36; field experiments 35;
genetically modified (GM) crops 35;
lack of protests against 37; "negative
knowledge" 35; politics of 35; social
solution to unidentified knowledge
gaps 35; support for in Finland 32;
unidentified knowledge gaps 35;
YIMBY ("yes-in-my-backyard") 32
Hirschman, Albert O. 47
historical culture 164
historical knowledge, of the Arctic 9
history: as resource 166; Russia's
institutional use of 168–169
Holt, Gemma 8, 178
homogeneity of communities in the
Arctic 3
human capital 72; in Lapland 123
human development 8
hydropower 42
hydropower plants 50; overspills 46

Iceland, issuance of climate bonds in 76
identity 4, 42; Sámi 146; Soviet 163, 167
imaginaries 176; resource-intensive
176–178
independence of Greenland 15, 16–18,
19–20, 22; actualization of 22;
discussing 23; elusiveness of 21;
entwinement with resource extraction
23–25
indicators 128
indigenous people: Arctic Council's
involvement of 107; Sámi 88
inequality 22–23
Institute of the North 57
institutionalization of a region, phases of
163–164
interdisciplinarity 5–6
International Arctic Science Committee
104
international mobility in Lapland 125–126
International Polar Year 2006–2008
initiative #299 57
International Polar Years (IPY) 133, 171
International Union for Conservation
of Nature (IUCN), "sustainable
development" 2
Interreg A 87, 88
Interreg V A Nord program 83–84, 88,
90, 96, 178; cultural resources of the
region 93–94; economic resources of
the region 92–93; Joint Operational
Program description 90; R&D in 92;
regional resources 91; Sámi people
94–95; social resources of the region
94–95; SWOT (Strengths, Weaknesses,
Opportunities, Threats) analysis 92;
tourism industry in the region 92–93
intersectionality 136–137, 139–140, 179
Inuit population of Greenland 21
investments 70; day trading 74; "green" 7,
71, 73, 79; in Greenland 19–20; SRIs
72–73; *see also* socially responsible
investments (SRIs)
investors 70; "socially responsible" 72–73
issuance of climate bonds 75; in Denmark
77–78; in Finland 78; in Iceland 76; in
Norway 76–77; in Russia 78–79; of
Sweden 77

Jääskö, Neeta 160
Jasanoff, Sheila 108, 178
Joint Operational Program description 90
Joy, Francis 8–9, 179

Katajamäki 95
Kemi River 43, 47, 49; flood management
 on 46, 50; hydropower plants 44;
 income from electricity production 47
Kemijärvi 33
Kemijoki River 6–7, 177; flood risk
 prevention 42
Kemijoki water reservoir 41
knowledge 8, 45, 72, 178; analyzing the
 treatment of in the AACA 109–112;
 ecological 106; ecological 107; and
 expertise 117–118; and happiness 36–37;
 historical 9; "negative" 35; non-Western
 112; policy transfer 87; policy-relevant
 108; scientific 107, 108; and second-
 wave feminism 136; "smart and arctic"
 118; traditional 146; usable 108–109
Kolari 84; development strategies 91;
 unemployment rates 85; *see also*
 Interreg V A Nord program
Korpelainen, Heikki 51
Kuokkanen 139
Kyoto Protocol 74

Lachesis 49
Laestadian Lutheran Church 28
language(s) 41; *meänkieli* 97; metaphors
 48; and names 48–49; Western 48; *see
 also* etymology
Lapin Kansa 50
Lapland 153, 179; Arctic expertise
 117–118; creativity in 123–125; human
 capital 123; international mobility in
 125–126; knowledge-based economy
 in 121–122, 123; Rovaniemi 125; smart
 specialization program 118; social
 dimensions of expertise in 119–120;
 well-being of its population 120, 121;
 see also Finland
Laruelle, Marlene 164
leadership index 124
legitimacy, of the Arctic Council 110–111
Lehtola, Veli-Pekka 148
Lempinen, Hanna 7, 9, 177
lifeline 48
lifespan 48
Lindholm, Maiju 153
Lindroth, Marjo 6, 176
linear thinking 42
linearity of Western languages 48
local needs versus global energy needs in
 the Arctic 59–60
Lohi, Markus 51

Lotvonen, Esko 51
Loviisa 32, 35

managing regional development 83
Mankala principle 31
Martello, M. L. 135
Marx, Karl 46; theory of value 50
material resources 5
Mathisen, S. R. 146
meänkieli 97
Medeiros, E. 87
member states of the Arctic Council 104
metal production 78–79
metaphors 41, 42, 48; bow of life 48;
 thread of life 49
mitigation 103
Moirai 49
Monument to the Frontier Guards of Arctic
 170
Monument to Waiting Woman in
 Murmansk 171
multiple knowledges, analyzing the
 treatment of in the AACA 109–112
Municipality Finance (MuniFin) 78
Muonio 84; development strategies 91;
 unemployment rates 85

names 48–49
nation building 16, 18; rationality of 24
natural power 45
natural resources 2, 4, 5, 8, 16–17, 42; in
 the European Arctic 72; in Finland 29;
 of Greenland 18
nature, in the Interreg V A Nord program
 region 91–92
"negative knowledge" 35
Noaiddit 144, 147, 148
noaidi drums 146
Nordic Sámi Convention of 2017 150
Nordregio 122
Northern Sea Route Administration 164,
 165
Norway 8; cultural appropriation of *Duodji*
 147; Interreg projects 87; Interreg V
 A Nord program 88; issuance of climate
 bonds in 76–77; *see also* Interreg V
 A Nord program; Sámi people
nuclear power production 28, 29, 38, 177;
 Akademik Lomonosov 33; domestic
 production of uranium 36; Fennovoima
 company 30, 30–31; Finland's history
 of 30; Finland's management of nuclear
 waste 34; in Finnish communities

29–33; fuel cycle 34; i 37; Mankala principle 31; Olkiluoto 3 31; protests 38; protests over 32; risks of living with 27; siting procedure 32–33; and unidentified knowledge gaps 35; worst-case scenario 27–28; YIMBY ("yes-in-my-backyard") 32
Nuorgam, Pia 157
Nuttal, Mark 16, 18, 19–20, 21, 131, 132
Nuuk 15, 16, 17, 19; discussion of Greenland's independence 23

oil and mineral strategy of Greenland 18, 19
Olkiluoto 3 31
Operation Derviš 168
opposition to resource projects 23–24
ordinem 49
Ounas River 44, 50
Ounasjoki 43
Our Common Future 2
overspills 46

Paasi, Anssi 163
Paltto, Aslak 152
Papanin, Ivan 168
Paris Agreement 63, 74
passe 148
Paukkajanvaara 34–35
Pello 84; development strategies 91; *see also* Interreg V A Nord program
performance, of green investments 73
Petrov, Andrey 123–124
Pielke, R. 114
Pike, A. 87
Pires, I. 87
Pirnes, Susanna 9, 179
Poirer, P. 165
polar exploration 9
Polar Record 131
polar research, exclusion of gender issues from 133–134
policy implications of the AACA 112
policy transfer 87
policymakers, on climate change 107
policy-relevant knowledge 108; theoretical basis for 108–109
political consumerism 37
politics: of adaptation 108, 178; *bios* 49; of extracting Greenland's resources 20; of flood management 51–52; and gender 132; gender and social sustainability discourse 134; of high-risk technology

35; power relations between Denmark and Greenland 23; and resources 42, 131; of the Russian Arctic 164–166; women's representation in 133
pollution 69
Posio, Rauno 153
Posiva 34
potentiality 44, 48; of the Kemi River 47; of rivers 49–50
power relations 41, 45; between Denmark and Greenland 23; and social order 45
predictability 45
privatization of flood insurance in Finland 47
Pro Hanhikivi protest 28
procedural dimension of sustainability 3
protests 38; Pro Hanhikivi 28
prudent development in the Arctic 61–63
Przybyslawski, Artur 41, 48
Putin Vladimir 165, 168
Pyhäjoki 6, 28, 35, 37–38; nuclear power production in 29, 30, 33; support for high-risk technology in 32

Radiation and Nuclear Safety Authority (STUK) 31
radioactivity: in the Barents region 34; disposal of nuclear waste 34, 37; Finland's management of nuclear waste 34; Fukushima 28, 33; risks of living with nuclear power plants 27, 33; and unidentified knowledge gaps 35; uranium as domestic product 36; *see also* uranium mining
radium 36
Raudanjoki 43
regere 42
Regional Council for Lapland 46
regional development 84, 86, 95, 97; endogenous 86, 87; exogenous 86, 97; in Finland 90; focus of 87; Interreg projects 87, 88; Interreg V A Nord program 88, 90–95; managing 83; resources for 83–84
regional energy security needs in the Arctic 58–60, 61, 62
regional expertise 119
Regional Potential Index (RPI) 122
renewable energy 33, 65; development of in the Arctic 59–60; in Finland 29; hydropower 42; *see also* energy resources; nuclear power production
replica drums 9

reproduction of fake Sámi drums in
Finland 152–153
reservare 45
reservoirs 45; flood management 51–52;
overspills 46
resource extraction 6, 16, 52; domestic
production of uranium 36; economics of
23; entwinement with Greenland's goal
of independence 23–25; in Greenland
17, 19, 20, 22; Greenland's political
discourse on 21; suspicious attitudes
towards 22–23; uranium 15
resource-intensive imaginaries 176–178
resource(s) 1, 4, 6, 7, 9, 15, 24, 42, 84,
131; Arctic 8; energy 7; in Finland
29; of Greenland 18; history as 166;
"human" 72; material 5; natural 4, 5;
in political discourse 42; for regional
development 83–84; social 131–132; as
sources of anxiety 19; theory of value 50
resurgere 42
risks of living with nuclear power plants
33
rivers 41, 42, 52; *course* 45; flood
management 42, 51; Kemi 43, 49;
Ounas 44, 50; potentiality of 49–50;
Tornio 46; units of 44
Rosatom 31
Rosmolodëž' 172
Rosner, V. 133
Rovaniemi 6–7, 43, 52; development of 51
Ruostetsaari, I. 37
Ruotsala, Anni-Helena 156
RUSAL 78–79
Russia 37, 163; *Akademik Lomonosov*
33; "Arctic!!!" 171; Arctic identity
building 164; Arctic identity of 172;
Arctic politics 172; Glavsevmorput
167–168; *Gosudarstvennyj sovet* 165;
Great Patriotic War 166, 170; *Gulag*
labor camps 168; Hamina Peace Treaty
84; identity building 166; institutional
use of history 168–169; involvement in
Finland's development 30–31; issuance
of climate bonds in 78–79; Kemi River
43; military activity in the Arctic 165;
Monument to the Frontier Guards of
Arctic 170; Monument to Waiting
Woman in Murmansk 171; monuments
170; Murmansk 170; national identity
166, 170; nationalism 164; Northern
culture 167; Northern Sea Route
Administration 165; Polar Explorer

Day 169; politics of the Russian Arctic
164–166; Rosmolodëž' 172; science
policy 169; St. Petersburg 170
Russian Arctic 9; as identity project
171–172
Russian Geographical Society 169
Russian North 167, 168
Rydin, Yvonne 128

saliency, assessing 111
Salla 2; development of 3–4; socially
unsustainable development in 3
Sámi Council 145
Sámi people 9, 88, 92, 94–95, 146, 179;
cultural appropriation, contributing
factors 147–148, 150; cultural
appropriation of *Duodji* 147; divination
drums 144; *Duodji* 145; *Noaiddit* 144,
147, 148; *passe* 148; protection of their
cultural heritage 153–155, 155–157;
sieidi 148; struggle for recognition 155
Scheffers, Johannes, *History of Lapland*
148
scientific exploration of the Arctic, Arctic
Monitoring and Assessment Programme
(AMAP) 105
scientific knowledge 107, 108; assessments
of the Arctic 107
second wave feminism 136
Sedov, Georgij 168
Sejersen, Frank 23
"self-destruction thesis" 47
Severikangas, Pertti 51
shamanism 9
sieidi 148
Silvén, Eva 153
Simo 33
Sinevaara-Niskanen, Heidi 8, 179
siting procedure for nuclear power plants
32–33
smartness 118, 119
Smith, Adam 47
social dimension of sustainable
development 3, 60, 61, 62
social imaginaries 9, 176, 178, 180;
resource-intensive 176–178
social order 42, 45
social resources 131–132
social responsibility 41
social sustainability 2, 4, 7, 8, 9, 41, 42,
164, 175, 176, 179–180; achieving
gender equality 136; of Arctic energy
development 57; expertise 8; in Finnish

Lapland 118; indicators of 128; of the
Kemijoki River 7
social values 8
socialist realism 170
socially responsible investments (SRIs) 7,
70, 71, 72–73, 79; ESG approach 73;
and financial products 74–76; purposes
71; societal benefits 73; *see also* climate
bonds
socially unsustainable development 3
societal development 7
Sotarauta, M. 82
sourse 42
Soviet Union: Agreement of Friendship,
Cooperation and Mutual Assistance with
Finland 30; collapse of 166; socialist
realism 170; and Soviet identity 163, 167
space, the Arctic as 167
SpareBank climate bond 76–77
stability, of green investments 73
stakeholder engagement 62
standpoint feminism 136
State Commission on the Arctic 164
state of pure contingency 45
stock exchange, trading on 74–75
Strauss-Mazzullo, Hannah 6, 176–177
Suomi, Kirsi 156
surgere 42
surrigere 42
sustainability 2, 3, 4, 6, 8, 9, 41, 140,
175–176; and adaptation 103–104;
of Arctic energy development 57;
cultural 166; inclusion of gender issues
in 135; indicators of 128; procedural
dimension 3; social dimension 3, 7, 60,
61, 62; and sustainable development 5;
see also cultural sustainability; social
sustainability
sustainable development 2, 3, 41, 70, 72,
83; Arctic Council 104; of the Arctic's
energy resources 60; versus prudent
development 61–63; social dimension
3; and sustainability 5
Sweden 7, 8; Hamina Peace Treaty 84;
institutional cross-border cooperation
88; Interreg projects 87; Interreg V
A Nord program 88; issuance of climate
bonds 77; Kiruna 88; *see also* Interreg V
A Nord program
SWOT (Strengths, Weaknesses,
Opportunities, Threats) analysis of the
Interreg V A Nord program 84, 90–95
symbolism, of divination drums
144–145, 151

Taking Wing conference 134
talent index 124
Talvivaara 36
Taylor, C. 180
technological development 59–60
technology; *see also* high-risk technology
Tennberg, Monica 8, 9, 133, 178–179
Teollisuuden Voima 34
theoretical basis for for policy-relevant
knowledge 108–109
theory of value 50
third wave feminism 136–137
Thisted, Kirsten 22–23
thread of life 49
Tödtling, F. 86
Tornio 84; municipal strategy 94; *see also*
Interreg V A Nord program
Tornio River 46
Tornio Valley 7
tourism 9; appropriation of Sámi culture
147–148, 150; consequences of using
Sámi cultural heritage as a resource
157, 159; culturally responsible 150;
fake divination drums as marketing
tool of 145; in the Interreg V A Nord
program region 92–93; Visit Arctic
Europe 153
trading on the stock exchange 74–75
Tulppo, Paula 7, 178

UNESCO Convention on the
Safeguarding of the Intangible
Cultural Heritage 150, 154
unidentified knowledge gaps 35
United Nations Conference on
Environment and Development 3
United Nations Convention on the Law of
the Sea (UNCLOS) 165
United Nations (UN) 3; Human
Development Index (HDI) 120
uranium, as domestic product 36
uranium mining 15, 17, 22, 28, 38; in
Greenland 23; Greenland's political
discourse on 21; Paukkajanvaara 34–35;
planning resource extraction projects in
Greenland 19
usable knowledge 108–109

Vasakronan real estate company 77
Visit Arctic Europe 153
Vladimirova, V. 133
Vodolazkin, Eugene, *The Aviator* 168
Vola, Joonas 6, 177

Water Framework Directive 45
water management 45
Water Scarcity and Drought
 Communication 45
well-being of Lapland's population 120, 121
Western languages 48
Weszkalnys, G. 21
Widdis, Emma 167
Wilson, Emma 23
Wilson, Gary 175
World Bank 79

worst-case scenario of living near nuclear
 power sites 27–28

Yakhina, Guzel, *Zuleikha Opens Her Eyes*
 168
YIMBY ("yes-in-my-backyard") 32
Ylitornio 84, 96; development strategies
 91; *see also* Interreg V A Nord program

zoë 49
"zones of sacrifice" 20

Printed in the United States
by Baker & Taylor Publisher Services

Printed in the United States
by Baker & Taylor Publisher Services